Veterinary Pathology and Veterinary Anatomy: 2nd Edition

Veterinary Pathology and Veterinary Anatomy: 2nd Edition

Guest Editors

Carmen Solcan
Gheorghe Solcan

MDPI

Basel • Beijing • Wuhan • Barcelona • Belgrade • Novi Sad • Cluj • Manchester

Guest Editors
Carmen Solcan
Department of Preclinics
Ion Ionescu de la Brad
University of Life Sciences
Iași
Romania

Gheorghe Solcan
Clinics Department
Iași University of Life Sciences
Ion Ionescu de la Brad
Iași
Romania

Editorial Office
MDPI AG
Grosspeteranlage 5
4052 Basel, Switzerland

This is a reprint of the Special Issue, published open access by the journal *Life* (ISSN 2075-1729), freely accessible at: https://www.mdpi.com/journal/life/special_issues/5YJ9L17412.

For citation purposes, cite each article independently as indicated on the article page online and as indicated below:

Lastname, A.A.; Lastname, B.B. Article Title. *Journal Name* **Year**, *Volume Number*, Page Range.

ISBN 978-3-7258-5873-6 (Hbk)
ISBN 978-3-7258-5874-3 (PDF)
https://doi.org/10.3390/books978-3-7258-5874-3

Contents

About the Editors

Carmen Solcan

Carmen Solcan is a Professor of Veterinary Histology and Embriology, Molecular Biology, at the University of Life Sciences Ion Ionescu de la Brad Iași; from 2016 to the present, she has been a doctoral supervisor in Veterinary Histology. In 1998, she completed her doctoral thesis in Veterinary Medicine at the University of Agricultural Sciences and Veterinary Medicine Ion Ionescu de la Brad in Iași. Her research fields include veterinary histology and immunohistochemistry (the reactivity of immune organs, and the biocompatibility or toxicity of some biomaterials). She has managed five research grants and participated as a member of 29 grant teams. She has published 286 works, of which 15 were monographs, textbooks, and practical guides; 273 were scientific articles, of which 90 were published in ISI journals, 183 were indexed in other International Databases.

From 2010 to 2011, she was a member of the National Council for Certification of Diplomas and University Certificates (CNATDCU), Veterinary Medicine commission. She has been awarded one prize from the Academy of Agricultural and Forestry Sciences, one prize from the General Association of Veterinarians from Romania and some other prizes in international scientific events (Inventica, Euroinvent).

Gheorghe Solcan

Gheorghe Solcan is a Professor, Department of Internal Medicine, Veterinary Dermatology and Toxicology, at the University of Life Sciences Ion Ionescu de la Brad Iași; from 2008 to the present, he has been a doctoral supervisor in Veterinary Medicine, with 24 completed doctoral theses, validated by CNATDCU.

In 1998, he completed his doctoral thesis in Veterinary Medicine at the University of Agricultural Sciences and Veterinary Medicine Ion Ionescu de la Brad in Iași. His research fields include veterinary internal medicine and veterinary toxicology. He has managed eight research grants and participated as a member of 27 grant teams. He has published 431 works, of which 25 are monographs, textbooks, and practical guides; 406 are scientific articles, of which 59 are in ISI journals with an impact factor, 23 are in ISI Proceedings, 13 are ISI abstracts, and 275 are indexed in other International Databases.

He has experience in management positions—between 2000-2004 and 2008-2012, he was scientific vice-dean of the Faculty of Veterinary Medicine in Iași; between 2004-2008 and 2008-2020, he was the administrative vice-dean of the Faculty of Veterinary Medicine in Iași.

From 2011 to the present, he has been a member of the National Council for Certification of Diplomas and University Certificates (CNATDCU), part of the Ministry of Education and Research, Veterinary Medicine commission, dealing with the validation of doctoral theses, habilitation theses and commissions, proposing quality standards for promotion in the academic career. He has been awarded two prizes from the Romanian Academy (in 2013 and 2018), three prizes from the Academy of Agricultural and Forestry Sciences, one prize from the General Association of Veterinarians from Romania and some other prizes in international scientific events (Inventica, Euroinvent). He was also awarded the title of Doctor Honoris Causa by the State Agricultural University of Moldova (Chișinău, 2019).

Preface

This Special Issue of *Life*, titled "Veterinary Pathology and Veterinary Anatomy", is dedicated to the study of veterinary disease and physiology and encompasses various aspects of pathology in domestic animals. It also focuses on new developments in animal anatomy, pathology, diagnosing and treating animal diseases. Original manuscripts, reviews and case reports covering various aspects of veterinary pathology were requested. Papers reporting new solutions to common problems and new descriptions of lesser-recognized conditions were particularly welcome. In addition, we encouraged papers detailing advanced diagnostic techniques, including the use of imagistic methods, histology and immunohistochemistry in diagnosis. Studies of diseases caused by industrial toxins, and mycotoxins (the current state of pathogenetic mechanisms, diagnosis, and means of prevention) were also requested.

Consequently, the evaluation process for this Special Issue primarily concentrated on the subjects described in the aforementioned objectives, all of which share new approaches in animal pathology as a central element, which is vital for the sustainability of animal welfare, their quality of life, and the safeguarding of human health and the environment, in alignment with the current One Health concept. The Guest Editors for the Special Issue personally reached out to numerous prospective authors. Ultimately, just 12 manuscripts were accepted following a comprehensive evaluation by two or more main reviewers selected for their specialization, particularly in animal anatomy or pathology. All submitted articles underwent improvements based on the feedback from the reviewers, and the final decision was made by one of the Guest Editors, based on the reviewers' assessments and evaluations. Therefore, as part of the Special Issue "Veterinary Anatomy and Veterinary Pathology-2nd edition", 12 articles covering various aspects of new trends in animal anatomy and pathology were published through a multidisciplinary approach. The results of the research demonstrated the progress in veterinary anatomy and pathology, along with the technical progress of society, to the benefit of both animal and human welfare.

So, anatomy benefits from modern 3D views of pieces in teaching and research, and these views also help the most modern methods of diagnosis and therapy: computed tomography, magnetic resonance imaging, endoscopic surgery, ultrasonograpy, thermography, etc. Additionally, some new electroencephalographic features in hepatic encephalopathy are presented. New sources of intoxication or a new rapid and sensitive method of ochratoxin detection are also included. The biocompatibility of some new biomaterials designated for repairing or reconstructing significant bone defects in human beings is also included.

This publication is geared towards veterinarians, researchers, graduate students, and learners in the fields of veterinary medicine and animal science, also serving as a valuable resource for the general public who are concerned with animal health and well-being.

Carmen Solcan and Gheorghe Solcan
Guest Editors

life

MDPI

Editorial

Special Issue "Veterinary Anatomy and Veterinary Pathology: 2nd Edition"

Gheorghe Solcan [1,*] and Carmen Solcan [2]

1 Internal Medicine Clinic, Faculty of Veterinary Medicine, Ion Ionescu de la Brad Iasi University of Life Sciences, 8 M. Sadoveanu Alley, 700489 Iasi, Romania

2 Department of Preclinics, Faculty of Veterinary Medicine, Ion Ionescu de la Brad Iasi University of Life Sciences, 700490 Iasi, Romania; carmen.solcan@iuls.ro

* Correspondence: gheorghe.solcan@iuls.ro

Anatomy is a basic science for both human and veterinary medicine. Veterinary anatomy faces new challenges. There is a need to understand current morphological fundamentals of animals through modern imagistic methods, such as computed tomography (CT), magnetic resonance imaging (MRI), ultrasonography, etc., but there is also a need to study new, exotic companions, or wild animals. Veterinary pathology is continuously progressing; in parallel with the progress of humankind, new strategies of diagnosis, prophylaxis, and therapy of animal diseases are being developed, to the benefit of both animals and human health and welfare.

This Special Issue, "Veterinary Anatomy and Veterinary Pathology: 2nd Edition", aimed to publish original research works, reviews, and case reports within these topics, highlighting their importance in the search for new diagnostic, prophylactic, and therapeutic strategies for animal diseases. This reprint consists of 12 articles that present recent insights and advancements in the fields of animal anatomy and pathology, authored by 77 researchers from various countries (Romania, Poland, Italy, South Korea, and USA), addressing contemporary issues through interdisciplinary and multidisciplinary methods, from the anatomy of a monkey to modern imagistic methods (computed tomography, endoscopy, ultrasonography, and electrophysiology) used in the diagnostic and treatment of various animal pathologies, or the biocompatibility of some new biomaterials on animal experimental models.

Mortonos et al. [1] provide a comprehensive overview of the anatomical features of the humerus in the African green monkey (*Chlorocebus sabaeus*), along with comparative and differential aspects of monkey osteology. The micromorphological findings offer valuable insights into this relatively under-researched primate, highlighting similarities with other primates regarding histological structures and detailing specific histomorphometric elements related to secondary periosteal bone formations (osteons).

Beginning from anatomy of healthy dog (*Canis familiaris*), Choi [2] succeeded to establish some correlations between hyperlipidemia-related diseases and thorax/thigh circumference ratio along with body condition score in dogs. According to author findings, triglycerides (TG) had a strong positive link with thigh circumference and a negative correlation with thorax/thigh value. While the thorax/thigh value was negatively correlated with the medial patella luxation (MPL) grade, total cholesterol (TC) was strongly positively correlated with thigh circumference. The Apolipoprotein E (ApoE) gene might be essential for thigh fat accumulation. Despite that, the thorax/thigh circumference ratio failed to be a new indicator comparable to the waist/hip circumference ratio in humans.

Computed Tomography (CT) is a modern imagistic method for the diagnosis of many internal diseases (especially those of the nervous system) in humans and companion animals, but it is cost-limitative for farm animals. Using CT scans to confirm the presence of cysts of *Coenurus cerebralis* in the brain of sheep with neurological symptoms [3], Olar et al. [4] have shown that plasma kynurenic acid levels, as determined by fluorescence spectroscopy [5], may be used as diagnostic indicators for sheep with chronic coenurosis, allowing practical application in early-stage diagnosis at the farm level.

MRI is an advanced imaging method, giving even more accurate results than CT in some pathologies, but the longer exposure period, under general anesthesia, leads to the risk of hypothermia [6]. Pavel et al. [7] have demonstrated that a hot water bottle placed beneath the abdomen and a blanket covering the entire body can prevent hypothermia in cats requiring a head MRI.

Endoscopy is another modern imagistic method for diagnostic and treatment of many diseases, both in human and in veterinary medicine, and endoscopic surgery became an increasingly popular, low invasive procedure. Laparoscopic nephrectomy is a procedure performed by Prządka et al. [8], and even if in previous studies, the ureter has been closed using mechanical clips or stitches [9], the authors have succeeded in demonstrating that it is feasible to close the renal vessels and ureter, in both cats and dogs, with vascular sealing tools, using advanced bipolar coagulation techniques. The ureter closure was proved by histopathological postoperatory exam.

Infrared thermography (IRT) is a method that evaluates peripheral blood flow and the heat emitted as a result, which is now utilized in human healthcare to noninvasively examine peripheral vascular issues like thrombosis, thromboembolisms, and various ischemic, inflammatory, or tumor-related conditions. The use of IRT for evaluating blood vessel issues in veterinary science is still in the early stages, with few research studies released [10]. Using IRT to assess the comparative thermal pattern of the thoracolumbar area in healthy dogs and dogs with acute intervertebral disc extrusion (IVDE), Zaha et al. [11] have identified notable variations in the thermal patterns observed in dogs suffering from IVDE, when compared to healthy dogs. They suggest using thermographic scans of the thoraco-lumbar region as a less invasive approach for diagnosing IVDE.

In the same line of non-invasive diagnostic techniques, Strichea et al. [12], using ultrasonography of the liver, identified multiple hyperechogenic strands diffusely dispersed throughout the parenchyma, giving it a "Swiss cheese" appearance, an aspect considered pathognomonic for a rare and severe condition in dogs—hepatocutaneous syndrome. So, authors recommend ultrasound assessment of the abdomen in dogs with dermatological symptoms compatible with hepatocutaneous syndrome.

Electroencephalography (EEG) offers a factual basis for diagnosing seizures and epilepsy; however, it is considered infrequently utilized in the field of veterinary neurology [13]. Ştefănescu et al. [14] have studied EEG in a pediatric dog with seizures consequent to portosystemic shunt (confirmed by ultrasonography and CT), concluding that the EEG recording displayed bilateral simultaneous bursts of three-phase waves, similar to non-convulsive status epilepticus. Authors recommend EEG as a complementary method of the diagnosis of portosystemic shunting, which might be useful in initiating the therapeutic protocol, before final confirmation using more laborious diagnostic methods.

Employment of animal models in translational studies for the development of bone tissue engineering and regenerative medicine is a relatively new research direction in which veterinary pathology is increasingly involved [15]. Investigating the osteogenic potential and toxicological tolerance of two original bioproducts, Ardelean et al. [16] proved the biocompatibility of the biomaterials for rats femoral bones. This study could open new research directions concerning the use of dental biomaterials in creating frameworks for the restoration of bone defects.

Corneal abrasion represents the predominant eye-related issue that arises during general anesthesia for nonocular surgeries. The pain experienced afterward can occasionally be more intense than the discomfort caused by the surgery itself [17], yet it frequently goes unrecognized in veterinary practice. Given that there is no established method for safeguarding the cornea during general anesthesia [17], Pavel et al. [18], investigating the lacrimal gland tear production in healthy sheep under general anesthesia, proved the protective effect of 1% hyaluronic acid ophthalmic gel, offering essential information to assist anesthesiologists in effectively overseeing the impact of general anesthesia related to safeguarding the cornea.

Toxicology is an important branch of veterinary pathology in continuous development. Ochratoxin (OTA) is very toxic and ranks among the most prevalent mycotoxins that contaminate animal feed globally [19], and the quantitative determination is laborious and expensive. Beia et al. [20] have developed a rapid and sensitive method for the quantification of OTA levels in maize, utilizing ultra-performance liquid chromatography coupled with fluorescence detection (UPLC-FLD). The method separates OTA from matrix interferences, thereby ensuring reliable identification at minimal levels, serving as a trustworthy method for monitoring mycotoxins to maintain feed security and safeguard both animal welfare and public health.

Copper is an essential trace element, playing an important role in many physiological processes: energy generation, immunity, body development, nervous system activity, and the formation of connective tissues, but it can become harmful in animals that are excessively exposed [21]. Domestic sheep are the species that show the highest sensitivity to copper toxicity, and to reduce the intoxication risk, inorganic salts were replaced in feed additives with organic ones. Even so, Pivariu et al. [21] firstly report a classical outbreak of chronic copper intoxication in sheep produced by feed with added copper bilysinate, but destinated for pigs, which are much more tolerant to copper. So, a mistaken change in destination of feed ingredients from one species to another is an important intoxication hazard.

In conclusion, this Special Issue has provided recent updates and important findings in veterinary anatomy and pathology, like new diagnostic and therapeutic strategies in various animal diseases, specifically advanced imaging techniques (CT in sheep coenurosis, MRI in cats, ultrasonography, and thermography), electroencephalography, modern surgery procedures (laparoscopic nephrectomy), biocompatibility of some new biomaterials, and prevention of some intoxications.

Author Contributions: Conceptualization, G.S. and C.S.; investigation, G.S. and C.S.; writing—original draft preparation, G.S. and C.S.; writing—review and editing, G.S.; visualization, G.S. and C.S. All authors have read and agreed to the published version of the manuscript.

Acknowledgments: We wish to convey our appreciation to all the authors who contributed manuscripts to the Special Issue of Life named "Veterinary Anatomy and Veterinary Pathology: 2nd Edition", the reviewers for their valuable feedback and recommendations, and the editorial team of Life.

Conflicts of Interest: The authors declare no conflicts of interest.

References

1. Martonos, C.O.; Gudea, A.I.; Little, W.B.; Stan, F.G.; Lațiu, C.; Bolfa, P.; Dezdrobitu, C.C. The Gross Anatomical and Histological Features of the Humerus in African Green Monkeys (*Chlorocebus sabaeus*) from Saint Kitts and Nevis, West Indies. *Life* **2024**, *14*, 1295. [CrossRef]
2. Choi, K. Correlation Between Hyperlipidemia-Related Diseases and Thorax/Thigh Circumference Ratio Along with Body Condition Score in Dogs Focusing on Molecular Mechanism: A Pilot Study and Literature Review. *Life* **2024**, *14*, 1441. [CrossRef]
3. Varcasia, A.; Tamponi, C.; Ahmed, F.; Cappai, M.G.; Porcu, F.; Mehmood, N.; Dessì, G.; Scala, A. *Taenia multiceps* coenurosis: A review. *Parasites Vectors* **2022**, *15*, 84. [CrossRef]
4. Olar, L.E.; Tomoiagă, V.D.; Mârza, S.M.; Papuc, I.; Beteg, I.F.; Peștean, P.C.; Musteață, M.; Lăcătuș, C.M.; Marica, R.; Pașca, P.M.; et al. Computed Tomography and Fluorescence Spectroscopy Blood Plasma Analysis Study for Kynurenic Acid as a Diagnostic Approach to Chronic Coenurosis in Sheep. *Life* **2024**, *14*, 1121. [CrossRef] [PubMed]
5. Shahzad, A.; Edetsberger, M.; Koehler, G. Fluorescence Spectroscopy: An Emerging Excellent Diagnostic Tool in Medical Sciences. *Appl. Spectrosc. Rev.* **2010**, *45*, 1–11. [CrossRef]
6. Daly, E.; Cardy, T.J.; McFadzean, W.; Taylor-Brown, F.E. Perianaesthetic Complications in Cats Undergoing MRI of the Brain. *J. Feline Med. Surg.* **2024**, *26*, 1098612X241275023. [CrossRef]
7. Pavel, R.; Fernoagă, C.; Neagu, A.G.; Costea, R. Effect of Hot Water Bottle and Cloth Blanket on Rectal Temperature During Magnetic Resonance Imaging of the Head in Cats Under General Anesthesia. *Life* **2024**, *14*, 1646. [CrossRef] [PubMed]
8. Prządka, P.; Liszka, B.; Suliga, K.; Antończyk, A.; Kiełbowicz, Z.; Kubiak-Nowak, D.; Dzimira, S.; Skalski, A.; Gąsior, L. Ureteral Closure Using Advanced Bipolar Vessel Sealing Devices During Laparoscopic Nephrectomy in Dogs and Cats: A Pilot Series of Clinical Cases. *Life* **2024**, *14*, 1681. [CrossRef]
9. Constant, D.L.; Florman, S.S.; Mendez, F.; Thomas, R.; Slakey, D.P. Use of the LigaSure Vessel Sealing Device in Laparoscopic Living-Donor Nephrectomy. *Transplantation* **2004**, *78*, 1661–1664. [CrossRef]
10. Mota-Rojas, D.; Ogi, A.; Villanueva-García, D.; Hernández-Ávalos, I.; Casas-Alvarado, A.; Domínguez-Oliva, A.; Lendez, P.; Ghezzi, M. Thermal Imaging as a Method to Indirectly Assess Peripheral Vascular Integrity and Tissue Viability in Veterinary Medicine: Animal Models and Clinical Applications. *Animals* **2024**, *14*, 14. [CrossRef]
11. Zaha, C.; Cărpinișan, L.; Schuszler, L.; Paula, N.; Căsălean, T.; Florea, T.; Cristina, V.; Sicoe, B.; Rujescu, C.; Dascălu, R. Thermographic Scan of the Thoracolumbar Area in Dogs with Acute Intervertebral Disc Extrusion (IVDE): A Retrospective Study. *Life* **2025**, *15*, 68. [CrossRef] [PubMed]
12. Strichea, A.-H.; Hreniuc, Ș.L.; Solcan, G. Non-Invasive Paraclinical Diagnosis of Hepatocutaneous Syndrome in a Dog. *Life* **2024**, *14*, 853. [CrossRef]
13. Luca, J.; McCarthy, S.; Parmentier, T.; Hazenfratz, M.; Linden, A.Z.; Gaitero, L.; James, F.M.K. Survey of Electroencephalography Usage and Techniques for Dogs. *Front. Vet. Sci.* **2023**, *10*, 1198134. [CrossRef]
14. Ștefănescu, R.A.; Boghian, V.; Solcan, G.; Codreanu, M.D.; Musteata, M. Electroencephalographic Features of Presumed Hepatic Encephalopathy in a Pediatric Dog with a Portosystemic Shunt—A Case Report. *Life* **2025**, *15*, 107. [CrossRef]
15. Muschler, G.F.; Raut, V.P.; Patterson, T.E.; Wenke, J.C.; Hollinger, J.O. The Design and Use of Animal Models for Translational Research in Bone Tissue Engineering and Regenerative Medicine. *Tissue Eng. Part B Rev.* **2010**, *16*, 123–145. [CrossRef]
16. Ardelean, A.I.; Mârza, S.M.; Marica, R.; Dragomir, M.F.; Rusu-Moldovan, A.O.; Moldovan, M.; Pașca, P.M.; Oana, L. Evaluation of Biocomposite Cements for Bone Defect Repair in Rat Models. *Life* **2024**, *14*, 1097. [CrossRef]
17. Grixti, A.; Sadri, M.; Watts, M.T. Corneal Protection during General Anesthesia for Nonocular Surgery. *Ocul. Surf.* **2013**, *11*, 109–118. [CrossRef]
18. Pavel, R.; Ene, I.; Costea, R. Exploring Lacrimal Gland Tear Production in Sheep under General Anesthesia: Examining the Potential Impact of Utilizing 1% Hyaluronic Acid Ophthalmic Gel. *Life* **2024**, *14*, 1038. [CrossRef]
19. Solcan, C.; Timofte, D.; Floristean, V.C.; Carter, S.D.; Solcan, G. Ultrastructural lesions and immunohistochemical analysis of Bcl-2 protein expression in the kidney of chickens with experimental ochratoxicosis. *Acta Vet. Hung.* **2013**, *61*, 344–353. [CrossRef]
20. Beia, S.-I.; Ion, V.A.; Gagniuc, E.; Bujor, O.-C.; Ivan, E.Ș.; Barbu, A.; Pitoiu, E.; Beia, V.E.; Bădulescu, L. Detection of Ochratoxin A in Maize and Its Potential Impact on Avian Pathology in Romanian Farms. *Life* **2024**, *14*, 1477. [CrossRef]
21. Pivariu, D.; Oros, A.N.; Tabaran, A.; Caloni, F.; Bolfa, P.; Nagy, A.-L. Chronic Copper Bilysinate Poisoning in Five Texel Sheep: A Case Report. *Life* **2024**, *14*, 1363. [CrossRef] [PubMed]

life

MDPI

Article

The Gross Anatomical and Histological Features of the Humerus in African Green Monkeys (*Chlorocebus sabaeus*) from Saint Kitts and Nevis, West Indies

Cristian Olimpiu Martonos [1], Alexandru Ion Gudea [2,*], William Brady Little [1], Florin Gheorghe Stan [2], Călin Lațiu [3], Pompei Bolfa [1] and Cristian Constantin Dezdrobitu [1]

1 Department of Biomedical Sciences, Ross University School of Veterinary Medicine, Basseterre P.O. Box 334, Saint Kitts and Nevis; cmartonos@rossvet.edu.kn (C.O.M.); brlittle@rossvet.edu.kn (W.B.L.); pbolfa@rossvet.edu.kn (P.B.); cdezdrobitu@rossvet.edu.kn (C.C.D.)
2 Department of Anatomy, Faculty of Veterinary Medicine, University of Agricultural Sciences and Veterinary Medicine Cluj-Napoca, Calea Mănăştur 3-5, 400372 Cluj-Napoca, Romania; florin.stan@usamvcluj.ro
3 Faculty of Animal Husbandry and Biotechnologies, University of Agricultural Sciences and Veterinary Medicine Cluj-Napoca, 400372 Cluj-Napoca, Romania; calin.latiu@usamvcluj.ro
* Correspondence: alexandru.gudea@usamvcluj.ro

Abstract: This paper presents a detailed gross description of all anatomical elements of the humerus in the African green monkey and provides comparative and differential elements on monkey osteology. The osteometric investigation adds value to the gross morphological investigation, adjoining metric data to the gross descriptive data set. An in-depth investigation of the microstructural aspects of the humeral bone tissue is provided, with qualitative and quantitative details and potential for diagnostic applications. Of the gross morphological elements described, several unique features specific to this species include the humeral head shape that presents with distinctive low convexity and caudal placement, the shape of the intertubercular groove, the less developed greater tubercle, and the disposition of the rotator cuff muscle insertion. Furthermore, the overall cranio-lateral curvature of the bone shaft was found to have a distinctive 154–155 degree of angulation of the diaphysis, and the well-developed medial epicondyle was observed with its distinctive medio-caudal retroflexion. The histological investigation was more indicative of a typical non-primate organization of the bone tissue, with laminar vascular and avascular structures combined with the presence of the secondary Haversian system involving a mixture of scattered and dense unorganized secondary osteonal structures. The histomorphometric investigation yielded metrical data for the secondary osteonal structures in terms of area (20,331 ± 5105 μm^2), perimeter, and vascular canal area (64,769 ± 257 μm^2).

Keywords: African green monkey; osteology; humerus; histology; osteometry; histomorphometry; Haversian systems; osteon; giant osteon; drifting osteon

1. Introduction

Taxonomically, the African green monkey, or AGM (*Chlorocebus sabaeus aethiopus*), is part of the *Cercopithecidae* subfamily [1], and together with the baboon and macaque is one of the most utilized species in non-human primate research [2,3]. The high importance of this species for biomedical research is confirmed by PubMed's high number of citations for studies that take advantage of this species as an animal model. This species is relatively easy to manage in captivity and is known for its reproductively prolific nature, making *Chlorocebus sabaeus aethiopus* a viable alternative for Rhesus monkeys [4–8]. The study of the skeletal system of primates in general, and the AGM in particular, is important for human medical doctors, biologists, and veterinarians because the locomotory apparatus may serve as the basis for research projects ranging from pure anatomical studies to integrated ecological approaches, and even for biomedical, forensic, pharmacodynamical, or diagnostic research. This paper aims to fill a significant gap in the descriptive documentation on the morphology of the skeleton of the AGM, likely driven by hyperfocus on previously mentioned species.

The humerus, which serves as the subject of our investigations as part of the forelimb skeleton, is the anatomic base of the arm area. This bone shows morphological features which help researchers to differentiate orthograde and pronograde primates [9]. These anatomical locomotor adaptations in primates play a fundamental role in human evolution and can help us understand our origins and transformation [10]. Primates' need to maintain social connectedness, explore new territories, forage to obtain adequate food, and escape from natural predators has made locomotion a crucial element for all animals and humans [11–13]. These types of activities are facilitated by the limbs, which support the body weight. The forelimb of quadrupedal mammals is more important than the hindlimb, because it supports a disproportionately larger amount of body weight during locomotion and is also involved in social and feeding behaviours [11]. Because of its topography, the humerus is an essential anatomic structure of the forelimb as its proximal end is a key component of the glenohumeral joint (*Articulatio humeri*) and provides insertion points for the rotator cuff muscles (*m. supraspinatus, m. infraspinatus, m. teres minor* and *m. subscapularis*). Also, the distal humerus is part of the elbow joint (*Articulatio cubiti*) and provides origin points for the flexor and extensor muscles of the carpus and manus [14]. The shaft of this bone provides insertion points for the flexor and extensor muscles of the shoulder and origin points for the flexor and extensor muscles of the elbow. The articular head of the humerus is the most important structure of the glenohumeral joint in humans, as it is an anatomic area which facilitates a large variety of movements: flexion, extension, abduction, adduction, and rotation [15].

The distal end of the humerus is made up of the humeral condyle (*Condylus humeri*), the trochlea (*Trochlea humeri*) medially, and the capitulum (*Capitulum humeri*) laterally [16]. The humeral condyle has vital importance in the movements and dynamics of the elbow, just as the humeral head does for the shoulder joint. Variations in locomotion between various primate species have created substantial evolutionary changes in the humerus in relation to elbow joint morphology [17].

In mammals, the function and style of locomotion will influence the developmental and morphological features of forelimb bones [11,18,19]. Each type of locomotion influences humeral core morphology, leading to differences in humeral curvature between terrestrial and arboreal monkeys secondary to habitual muscle loads [19]. Even if the AGM is well adapted for ground (terrestrial) life, it has also developed in adaptation to some arboreal locomotion behaviour. This animal commonly utilizes terrestrial locomotion for feeding and resting, while arboreal movement is frequently observed during social behaviour and when they observe imminent danger [20].

The microstructural differences in bone histology have been a topic of research for a long time. Although many structural details have been described for numerous animal species [21–29], more recently, a series of studies investigated the differences in histologic bone structure between humans, primates, and other species [30–33]. There is a substantial body of research documenting forensic studies and anthropological and legal medical investigations which have focused on microscopic and histomorphometric details [34–38] in an attempt to differentiate and understand species and evolution at the microstructural level [39–43]. Other research utilizes these details in an attempt to explain different pathological expressions of disease, such as ageing, osteoporosis, or metabolic diseases [7,8,44,45].

This area of interest focuses on the basic organization of the mammalian compact bone, with an emphasis on periostal and endosteal areas. The Haversian system describes the basic structural units of the compact bone [29,33], even though in some mammalian species these units are absent [7,8,21,32].

The Haversian system focusses on osteonal units (secondary osteons) as the primary descriptive elements of osseous tissue. Each secondary osteon contains a central vascular canal (Haversian canal) surrounded by several bony laminae, each demarcated by a cement line [21,29,46]. In contrast, primary osteons are represented by vascular canals surrounded by relatively few concentric laminae, without the visible concentric lamellae created by the cement lines described in secondary (mature) osteons. Most works rely on this system and assess different arrangements listed in the table below based on the classification system of De Riqueles [22,23,28,30].

Primary (periosteal) bone type:

1.a Lamellar non-vascular canals;
1.b Lamellar simple (primary) vascular canals;
  1.b.1 Longitudinal;
  1.b.2 Circular;
  1.b.3 Reticular;
  1.b.4 Radial;
1.c Lamellar with primary osteons;
  1.c.1 Longitudinal primary osteons;
  1.c.2 Longitudinal primary osteons with radial canals;
  1.c.3 Longitudinal primary osteons with reticular canals;
  1.c.4 Longitudinal primary osteons and radial simple vascular canals;
  1.c.5 Longitudinal primary osteons in circular rows;
1.d Fibrous non-vascular bone;
1.e Fibrous bone with simple (primary) vascular canals;
  1.e.1 Longitudinal;
  1.e.2 Circular;
  1.e.3 Reticular;
  1.e.4 Radial;
1.f Fibrous bone with primary osteons (fibrolamellar complex);
  1.f.1 Laminal;
  1.f.2 Plexiform;
  1.f.3 Reticular;
  1.f.4 Radial;
  1.f.5 Laminar/plexiform with longitudinal primary osteons;
    1.f.5.a In circular rows;
    1.f.5.b In a band;
  1.f.6 Radial with primary osteons in radial rows;
  1.f.7 Longitudinal primary osteons;
  1.f.8 Longitudinal primary osteons in circular rows;
  1.f/1a-Pseudo-fibrolamellar complex.

Secondary periosteal bone types:

2.a.1 Scattered osteons;
  2.a.1.a Scattered osteons with no organization;
  2.a.1.b Circular rows of scattered osteons;
2.a.2 Dense osteons;
  2.a.2.a Dense osteons with no organization;
  2.a.2.b Circular rows of dense osteons.

To the best of our knowledge, the gross anatomy and histological features of the humerus in AGMs from Saint Kitts and Nevis have not yet been adequately investigated. Therefore, this study aims to provide an accurate and complete set of anatomical, structural, metrical, and functional information regarding the humerus in *Chlorocebus sabaeus* monkey. The outcome of this work completes the information provided by previous studies [29] related to the morphological characteristics of the scapula in AGMs [3]. These data will be useful for general practitioners, researchers, and students through enhancement of the anatomical knowledge of this species' skeletal macro- and microanatomy.

2. Materials and Methods

2.1. Animal Material

The biological material for this study utilized five complete skeletons which were part of a private collection hosted on Saint Kitts Island. The owner specifically permitted the study of these specimens in the anatomy laboratory of the Ross University School of Veterinary Medicine, Saint Kitts and Nevis, Basseterre. This study complied with the IACUC regulations (TSU10.27.2023 CM) from the Ross University Institutional Animal Care and Use Committee.

Examination of dentition indicated that all specimens were from adult animals, three males (k930, k945 and k920) and two females (a438 and v585). The right and left humeri from each specimen were collected for morphological and morphometrical analysis. Each humerus was carefully evaluated for its gross morphology, and the most important anatomical and metrical features were assessed and described.

2.2. Gross Anatomical and Osteometric Investigation

The humerus of each specimen was photographed in standard orientations to capture each anatomical segment of the bone from multiple perspectives. Precise measurements were taken via a scale placed near the bone in each image (Figure 1). The images were obtained with a DSLR Canon EOS 90D and were later processed with the Adobe Photoshop® program for fine contrast and background adding. Other adjustments and assessments were made with GIMP® and ImageJ® 1.54J software using the measuring tools add-ons available [47,48].

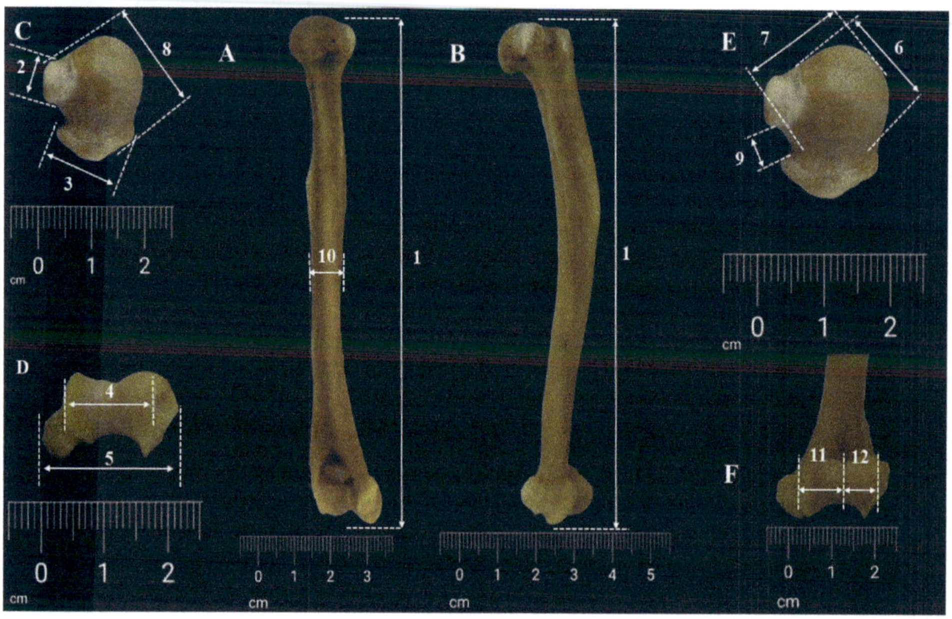

Figure 1. Measurements of humerus (**A**,**B**). Whole-bone measurements (**C**,**E**), proximal extremity measurements (**D**,**F**), and distal extremity measurements. 1. ML/MLH, maximum humeral length; 2. lesser tuberosity diameter; 3. greater tuberosity diameter; 4. WDASL, width of surface of lower end; 5. BED, biepicondylar distance; 6. TDH, transverse diameter of humeral head; 7. VDH, vertical diameter of humeral head; 8. WDU, width of upper end; 9. VDG, width of bicipital groove; 10. TDMS, transverse midshaft diameter; 11. TWD, width of trochlea; 12. CWD, width of capitulum.

The anatomical terminology used by the present study is in accordance with the sixth edition of Nomina Anatomica Veterinaria 2017 [49].

Using the data reported earlier in the literature (Figure 1), the following measurements were performed on the humerus bone from *Chlorocebusus sabaeus aethiopus* [20,50,51].

- ML/MLH—the total length of the humerus (as the minimum distance from the most proximal point on the head to the most distal point on the trochlea), also listed in other sources as the "functional length of the humerus";
- WDU—the width of the upper end (as the maximum transversal distance between the elements of the proximal end);
- VDH—the vertical diameter of the head (direct distance between the highest and the lowest point on the articular margin of the head);
- LTD—lesser tuberosity diameter (the maximum diameter of the lesser tuberosity measured at a right angle to the proximal shaft axis);
- GTD—greater tuberosity diameter (the maximum diameter of the greater tuberosity measured at a right angle to the proximal shaft axis);
- TDH—the transversal diameter of the head;
- VDG—the width of the bicipital groove;
- TDMS—transverse midshaft diameter;
- BED—biepicondylar distance;
- TWD—trochlear width;
- CWD—condylar width/capitulum width;
- WDASL—the width of the surface of the lower end.

The measurements were taken using the digital images presenting the physical scale most frequently. Some other measurements were taken using digital sliding callipers (0.01 mm).

2.3. Histological Technique

For the histological investigations, three cross-sections from three areas of interest were assessed (Figure 2) in the area of the surgical neck of the humerus, H50 and H40. H40 is the cross-sectional area measured at 40% of the biomechanical humerus length from the distal end, while H50 is the cross-section measured at 50% of the biomechanical humerus length, as suggested by previous researchers [52]. Decalcified bone samples (decalcification by soaking in approx. 125 mL of DeltaFORM™ decalcifying solution for 6 days) were included and processed through regular histological procedures. Five-micrometre slices were stained with "Toluidine Blue" comprising the following stages: deparaffinization and rehydration of the slides, staining with 0.4% Toluidine Blue solution for 10 min, several rounds of rinsing, and counterstain with 0.02% Fast Green solution followed by repeated rinsing and dehydration in ethanol and final clearing [53]. The histological images were evaluated qualitatively using an Olympus BX (Olympus Corporation of the Americas, Center Valley, PA, USA) conventional light microscope, according to the classification system of bone structure types adapted from Riqueles' system [22,23,54,55], marking the intensity of different types of bone with + signs in an incremental scale (+/++/+++). This style of bone description follows the classification system proposed by Enlow–Brown, focusing on tissue organized near the periosteum and endosteum as well as in the mid-region of the bone [21].

Image acquisition was performed with an Olympus DP 26 (Olympus Corp, Hamburg, Germany) digital camera and the cellSense Standard® 3.0 image analysis software. The ImageJ® 1.54J software application [47] was utilized for morphometrical analysis [47]. When necessary, the auto-contrast feature in the software was utilized to attain improved visualization. The Find Edges tool was also employed to highlight the cement lines and vascular canals. The dimensional features of the units—the vascular canals and the area of the secondary osteons—were measured using the elliptic tool features, which calculate a series of metric values [56]. We manually traced the boundary of intact secondary osteons and their Haversian canals in the images to apply the Measure tool within

ImageJ®. Data were collected using Excel and were later investigated with more advanced statistical tools. Primary statistical analysis was accomplished with Excel® version 2409, and more advanced statistics were attained utilizing the Statistics Calculator™ (https://www.socscistatistics.com/tests/) and OnlineStats™ (https://astatsa.com/ accessed July–September 2024) including the Kolmogorov–Smirnov test for normality, one-way ANOVA, Tukey, Sheffee, etc.

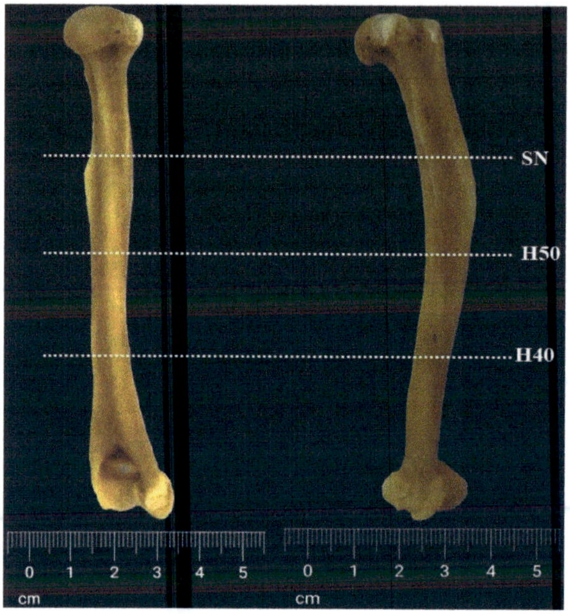

Figure 2. The divisions of the humeral shaft. SN—surgical neck area, H50—50% of the biomechanical length, and H40—40% of the biomechanical length.

3. Results

3.1. Macroanatomical and Metrical Features

These investigations revealed that the humerus of the AGM has a similar morphology to other mammals. The bone can be divided into three segments with two epiphyses (proximal and distal) and the shaft or the diaphysis, with each epiphysis possessing articular and non-articular elements.

From an overall perspective, we noticed that the curvature of the long axis of the bone was most accentuated at the proximal half of the shaft. This curvature was relatively enhanced in its cranial proximal direction. The angulation was evaluated and measured by placing the bone in anteroposterior presentation onto the measuring board, with the visual perspective focused on the olecranon fossa.

The surgical neck area was determined relatively close to the deltoid tuberosity, below the crest of the greater tuberosity (Figure 3). The images were all stretched to the upper and lower lines (despite metrical differences maintaining the overall shape and presentation). The upper and lower lines served as reference points for the place where the proximal and distal ends of the ImageJ ® 1.54J [47] angle tool were placed, and we used the pivot function to orient the location where the midline divides the diaphysis in half. The measurements taken (Table 1) indicate an average value for the angle of 155° (+/−2.5°). Due to a very limited number of measurements, no distinction between sexes was attempted. There was an overall length difference (ML) between males and females of approximately 8–10%, with an average value for males of 131.92 mm and 120.65 mm for females.

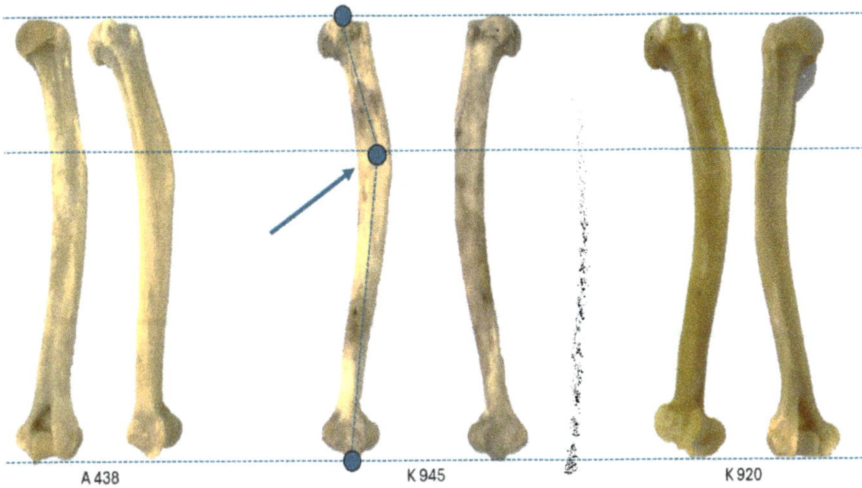

Figure 3. The angulations of the humerus in the standard presentation. The points of reference for the angle tools are marked by the blue dots (surgical neck area).

Table 1. The values for the assessed angulation of the diaphysis.

Specimen	R	L
a438	158	156
k945	153	157
k920	156	153

The proximal epiphysis (Figure 4) possesses an articular element (articular head of the humerus, *Caput humeri*) (Figure 4B,C) and non-articular structures (the greater tubercle, *Tuberculum majus*, and the lesser tubercle, *Tuberculum minus*) (Figure 4A–C). Between the last two structures, a U-shaped groove, the intertubercular groove (*Sulcus intertubercularis*) or the bicipital groove, is visible (Figure 4A).

The head of the humerus (Figure 4B,C) was the only articular structure of the proximal epiphysis, and in the AGM, this area accommodates low convexity and a flattened aspect with caudal orientation. The caudal position and the oval shape of the articular surface illustrate the participation of this structure as an important component in a high-mobility joint, the glenohumeral joint (*Articulatio humeri*). Caudo-distally, this structure continued with the neck of the humerus (*Collum humeri*) (Figure 4B,C), and cranially, it was continued by the intertubercular groove (Figure 4A).

The neck of the humerus was identified in all specimens as visible only on the caudal side of the proximal epiphysis. This location was to be the point of fusion between the articular and non-articular structures with the humeral body (*Corpus humeri*).

In AGMs, the non-articular structures of the proximal epiphysis have an important role as insertion points for the rotator cuff muscles. The morphological aspects of these structures can provide important information about the functions of these muscles.

In *Chlorocebus sabaeus aethiopus*, the greater tubercle (*Tuberculum majus*) has two segments (Figure 4A): the cranial segment (*Pars cranialis*), located cranially in regard to the humeral head, and a well-developed caudo-lateral segment (*Pars caudalis*) located on the lateral aspect of the proximal epiphysis. Although the greater tubercle is well developed, its height does not exceed the humeral head's height in the studied specimens. A careful examination of this tuberosity showed some muscular insertion points.

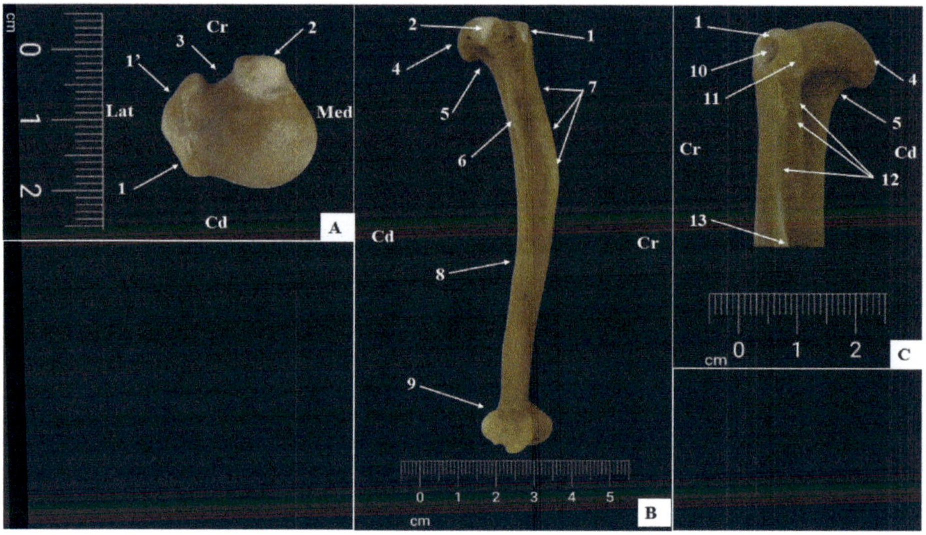

Figure 4. (**A**) Proximal extremity, (**B**) lateral perspective of bone shaft, and (**C**) detailed perspective of proximal–lateral part of bone. 1. Greater tubercle, caudo-lateral segment, 1′. greater tubercle, cranial segment, 2. lesser tubercle, 3. intertubercular groove, 4. humeral head, 5. anatomical neck of humerus, 6. teres major tuberosity, 7. crest of greater tubercle, 8. humeral diaphysis, caudal aspect, 9. humeral condyle, 10. fossa of bursa subtendinea m. infraspinati 11. insertion point for teres minor muscle, 12. tricipital line, and 13. deltoid tuberosity.

The most important and evident one was located on the lateral aspect of the caudo-lateral segment of the greater tubercle and serves as the insertion point for the infraspinate muscle (*M. infraspinatus*) and teres minor muscle (*M. teres minor*) (Figure 4C). The dorsal aspect of the greater tubercle serves as the insertion point for the supraspinate muscle (*M. supraspinatus*).

On the lateral aspect of the greater tubercle, we identified a small excavation which accommodates a subtendinous bursa (*Bursa subtendinea m. infraspinati*) (Figure 4C) located between the insertion tendon of the infraspinatus muscle and the proximo-lateral aspect of the greater tubercle of the humerus.

The lesser tubercle (*Tuberculum minus*) (Figure 4A,B) was smaller than the greater tubercle, was located on the medial aspect of the proximal extremity, and did not exceed the height of the humeral head. It also serves as the insertion point for the last of the rotator cuff muscles, the subscapular muscle (*M. subscapularis*).

The humeral diaphysis (*Corpus humeri*) (Figure 4B) is well developed and robust. The proximal third of the diaphysis has a curved contour with cranial concavity and caudal convexity. The presence of some bony crests, together with several bony prominences, made the identification of the four surfaces—cranial surface (*Facies cranialis*), caudal surface (*Facies caudalis*), medial surface (*Facies medialis*), and lateral surface (*Facies lateralis*)—extremely easy. The lateral surface (*Facies lateralis*) was the direct continuation of the greater tubercle of the humerus and allowed for elongated deltoid tuberosity (*Tuberositas deltoidea*) (Figures 4C and 5B). Between the base of the deltoid tuberosity and the caudal aspect of the greater tubercle, a fine bony crest was observed which was identified as the tricipital line (*Linea m. tricipitis*) (Figure 4C). The brachial groove (*Sulcus m. brachialis*) could be identified on the caudal surface of the proximal third and on the latero-cranial surface of the distal third of the humerus.

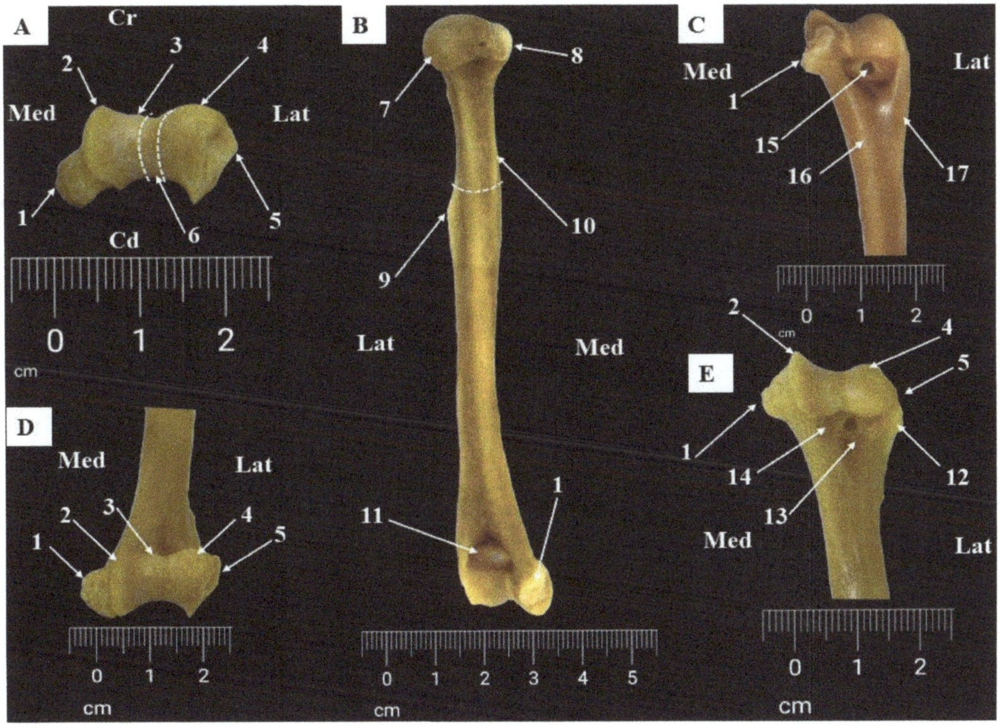

Figure 5. Morphological features of humerus bone shaft (**B**) and distal end (**A**,**C**–**E**). 1. Medial epicondyle, 2. medial trochlear keel, 3. lateral trochlear keel, 4. humeral capitulum, 5. lateral epicondyle, 6. zona conoidea, 7. humeral head, 8. lesser tubercle, 9. deltoid tuberosity, 10. teres major tuberosity, 11. olecranon fossa, 12. tail of lateral epicondyle, 13. radial fossa, 14. coronoid fossa, 15. supratrochlear foramen 16. Medial epicondyle, and 17. Lateral epicondyle Dotted line—humeral surgical neck area.

On the medial aspect of the proximal third of the humeral shaft, the teres major tuberosity (*Tuberositas teres major*) (Figures 4B and 5B) was evaluated. The crest of the lesser tubercle (*Crista tuberculi minoris*) (Figure 4B) was marked by an exceptionally fine bony line between the teres major tuberosity and the lesser tubercle of the humerus. The distal third of the medial surface revealed the presence of a distal nutrient foramen of the humeral diaphysis.

The cranial surface showed an evident bony crest which represents the distal continuation of the cranial segment of the greater tubercle of the humerus known as the crest of the greater tubercle (*Crista humeri*) (Figure 4B).

The caudal surface had a rounded appearance in latero-medial direction and distally ended with a pronounced excavation, the olecranon fossa (*Fossa olecrani*) (Figure 5B,C).

The distal end of the bone, the condyle of the humerus (Condylus humeri) (Figure 5E), contains articular and non-articular structures. The articular surface was represented by a medial trochlea (*Trochlea humeri*) and a lateral capitulum (*Capitulum humeri*) (Figure 5A–E). Specific to the AGM, the medial trochlear keel (pronounced medial flange of the trochlea) was very well developed compared with the lateral trochlear keel [57], which was very fine. Also, the medial keel surrounded the posterior aspect of the medial epicondyle and stopped near the olecranon fossa. Because of the small size of the medial trochlear keel, the zona conoidea (Figure 5A) was challenging to identify in our specimens. The cranio-lateral edge of the capitulum has a dorsal continuation known as the capitulum's tail (Figure 5E).

The non-articular structures noted were represented by the medial and lateral epicondyles and some fossae located on the cranial and caudal surfaces of the distal end of humeral diaphysis.

In *Chlorocebus sabaeus aethiopus*, the medial epicondyle (*Epicondylus medialis*) (Figure 5A,C–E) was noted to be particularly very well developed and pulled caudally compared to the lateral epicondyle (*Epicondylus lateralis*) (Figure 5A,B,D,E). The medial and lateral supracondylar crests (*Crista supracondylaris medialis* and *Crista supracondylaris lateralis*) (Figure 5C) continued beyond the epicondyles on the caudal aspect of the distal end of the humerus and were noted to be narrow and fine linearly.

The olecranon fossa (*Fossa olecrani*) (Figure 5B) was deep and in three specimens had direct communication with the radial fossa (*Fossa radialis*) (Figure 5E) and the supratrochlear foramen (*Foramen supratrochleare*) (Figure 5C). On the cranial aspect of the distal epiphysis, medially located, in relation with the fossa radialis and above with the humeral trochlea, a small coronoid fossa (*Fossa coronoidea*) (Figure 5E) was also observed.

The collected metrical data are listed in the table below (Table 2).

Table 2. Measurements on humerus (mm).

Specimen	**A438**	**A438**	**V585**	**V585**	**K920**	**K920**	**K930**	**K930**	**K945**	**K945**
Sex	Female	Female	Female	Female	Male	Male	Male	Male	Male	Male
Side	R	L	L	R	R	L	L	R	L	R
ML/MLH	125	131	123.6	115	125	128	140	137.9	140.89	138.1
WDU	19.5	19.44	19.31	18.8	21	20.6	21	22.3	22.9	22.9
VDH	13.59	13.14	13.13	12.42	15.2	13.66	15.82	15.84	13.98	14.91
LTD	5.77	6.33	8.68	7.35	7.41	6.13	8.99	9.24	9.05	9.85
GTD	12.29	11.96	12.42	13.14	14.44	13.6	13.85	14.88	13.21	13.1
VDG	4.21	4.49	4.35	4.47	5.09	4.5	5.6	5.8	4.93	5.09
TDMS	10.2	10.07	8.9	9.1	10.3	9.3	9.47	11	11.8	10.5
BED	21.4	18.9	21.5	22	21	21.3	25	27	23.17	23.1
TWD	8.27	8.03	9.64	9.65	7.76	8.99	9.37	9.5	10.3	10.6
CWD	6.2	6.37	5.31	6.4	7.27	7.8	8	7.14	6.59	5.5
WDASL	15.9	14.58	15.5	16.8	14.95	15.9	19	17.39	17.8	16.7

3.2. Micromorphological Evaluation of the Bone Tissue

3.2.1. Qualitative Evaluation of the Bone Histology

Each of the presented slides was first qualitatively assessed, specifically, the presence and the occurrence of the observed arrangements of the designated types of bone according to previously documented protocols (+++/++/+/) (Table 3).

Each of the samples observed originated from the previously three designated areas (surgical neck area, middle of the humerus H50, and at the area designated as the 40% biomechanical length of the bone—H40). For each of the main sectors, the circular section was divided into 8-10 radial sections and numbered accordingly.

As a general observation, several lamellar arrangements were visible within the core of the bone tissue. Lamellar bone and primary osteons were noted, but there was no visible pattern of arrangement according to subtypes.

Table 3. Assessment of bone tissue types.

	Primary Bone Types						Secondary Bone Types	
	1a	**1b**	**1c**	**1d**	**1e**	**1f**	**2a1**	**2a2**
H40/86	+++1a	+++1b +1b1	+1c1				++2a1a	+++2a2a
H40/88	+++1a	+++1b ++1b1					+2a1a	
H40/89	+++1a	+++1b ++1b1	++1c1				+2a1a	
H40/90	+++1a	+++1b +1b1	+1c1			+1f	+2a1a	
H40/91	+++1a	+++1b ++1b1					+2a1a	
H40/92	+++1a	+++1b	+1c1 +1c4			+1f	+2a1a	
H40/93	+++1a	++1b	+1c1				++2a1a	++2a2a
H40/94	++1a	++1b	++1c1					+++2a2a
H40/95	+++1a	+1b	++1c1				+2a1a	
H40/96	+++1a	+++1b +++1b1	++1c1 +1c2 +1c4			+1f7	+2a1a	+++2a2a
H40/97	+++1a	+++1b +++1b1	+1c1				+2a1a	+++2a2a
H50/58	++1a		+++1c1 +1c2				+2a1b	+++2a2a
H50/59	++1a	+++1b +++1b1	++1c5			++1f5 +1f3 ++1f7+1f2		+++2a2a
H50/60	++1a	+1b1	+++1c1			++1f7		+++2a2a
H50/61	+1a	+1b1	+++1c1				++2a1b	++2a2a
H50/62	++1a	++1b1	+++1c1				++2a1b	++2a2a
H50/63	+1a	+1b1	+++1c1					+++2a2a
H50/64	+1a	+1b1	+1c1			+1f1		+++2a2a
H50/65	+1a	+1b1	+1c1			+1f1		+++2a2a
H50/66	++1a	+1b1	++1c1				++2a1a	
H50/67	+1a	+1b1					++2a1a	
H50/69	++1a	++1b1	++1c1 +1c2					+2a1a
H50/70	++1a	++1b1	+1c1					++2a1a
H50/72	+1a	++1b1	++1c1			+1f1/1a-c		+2a1a
SN/29	++1a	+1b1	+1c1					+2a1a
SN/30	++1a	+1b1						+2a1a
SN/31	++1a	+1b						+++2a2a
SN/32	+1a	++1b						+++2a2a
SN/33	+1a							+++2a2a
SN/34	++1a	++1b1						+++2a2a
SN/35	+++1a	++1b1	+1c1			+1f1/1ac	+2a1a	++2a2a
SN/36	+++1a	+1b1				+1f1/1ac	++2a1a	+2a2a

Table 3. *Cont.*

	Primary Bone Types						Secondary Bone Types	
	1a	**1b**	**1c**	**1d**	**1e**	**1f**	**2a1**	**2a2**
SN/37	+++1a	+1b1					++2a1a	+2a2a
SN/38	+1a	+1b1	+1c1			+1f1	++2a1a	+2a2a
SN/39	+++1a	+1b1	++1c1				+2a1a	
SN/42	+++1a	+1b1	++1c1				++2a1a	+2a2a

The occurrence of areas with dense second osteonal arrangements was a common finding among the examined specimens (Figure 6). The scattered osteonal arrangement was not as commonly observed; however, it was documented several times in the examined slides. These findings of different bony arrangements observed with some lamellar vascular longitudinal canals should be mentioned, as the reticular Haversian vascular arrangement was not observed. Very rarely, the fibrous primary bone was noticed and was mostly present in very confined areas of our slides in the form of a pseudo-laminar arrangement closely related to primary osteonal units.

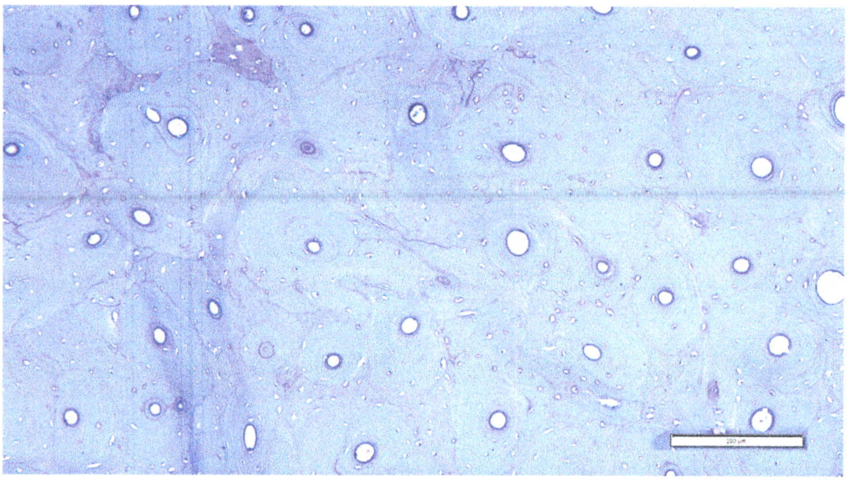

Figure 6. The osteonal units with no clear arrangement (surgical neck area), SN/33.

It should also be noted that the existence of some large super-osteons was visible intermittently, mostly notable at the level of the surgical neck area (SN) (Table 2).

In some of the examined visual fields, drifting osteons (also known as "waltzing osteons") were noted as present (Figure 7) [58–60].

Other special features noted through observations may be secondary to the peculiar distributions and prevalence of fine differences among the studied specimens. This is especially true considering the slight change in arrangements noted as we progressed from H40 to the surgical neck (SN).

- The lamellar avascular of primary bone [32] was prevailing and more intensely noted in the more distal part of the bone (H40), slightly reducing its importance in the midpart of the bone (H50), and seemed to regain its quota in the surgical neck area (SN).
- The lamellar simple vascular structure had a more intense presence in the H40 area, decreasing in intensity in H50, but not regaining its full occurrence in the SN area (maybe less intensely compared to the lamellar avascular component).

- The lamellar primary bone with primary osteons, mostly in longitudinal arrangements, was noticed in all segments of the bone, but the intensity and prevalence of this component were much rarer than those of the previous ones, with a notable presence in H50 when compared to the other two regions.
- The dense secondary osteonal arrangement (Figure 8) [22,23] was constantly present in most of the studied slices, usually showing an unclear organization pattern. In some situations, the scattered osteonal arrangement was visible. The density of secondary osteons seemed to occur more in the H50 and neck area, but this assessment is quite subjective due to the constant presence of these areas throughout the studied samples.

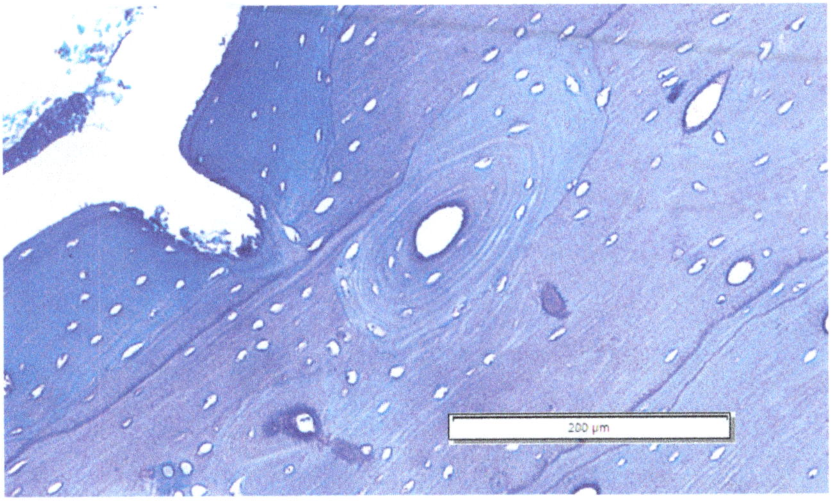

Figure 7. The drifting osteon: a rarely seen drifting osteonic unit situated close to the endosteal area of the A40 section.

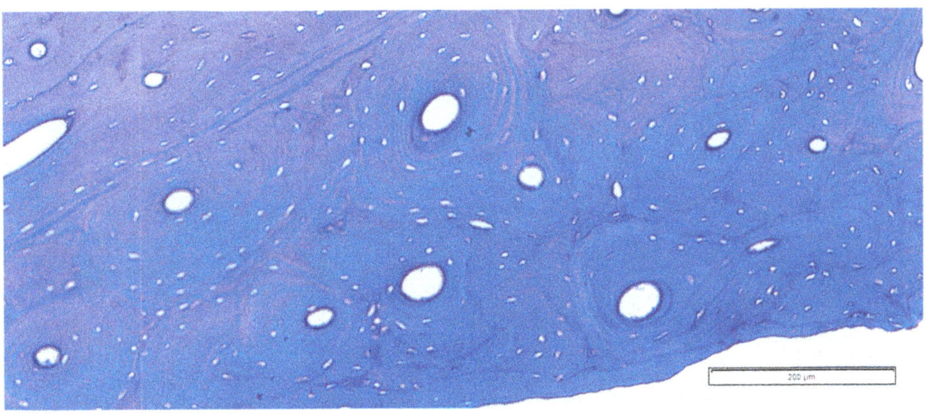

Figure 8. Area of H50/62 with dense secondary osteonal units.

Osteon banding (as a rare linear arrangement of secondary osteons) (Figure 9) was also noticed, but it seems to have a very low frequency and intensity within the examined specimens [58,59,61,62].

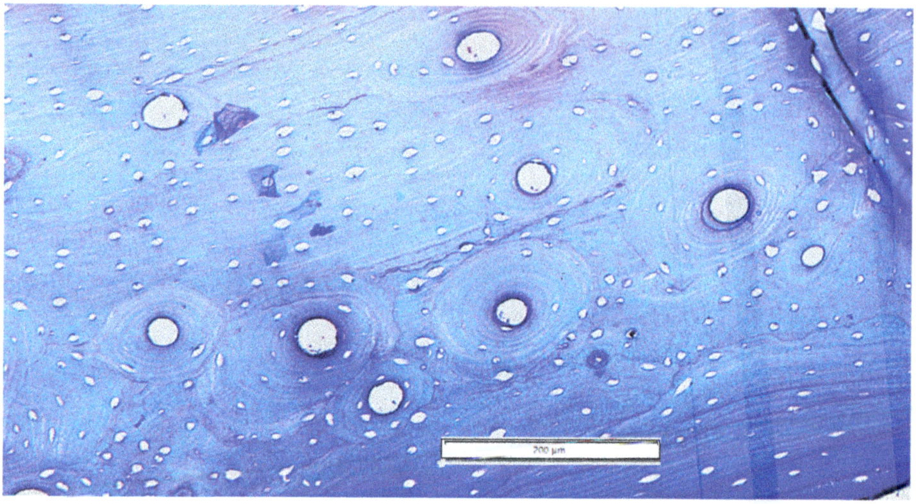

Figure 9. Osteon banding in H40/97 area.

3.2.2. Histometrical Investigation

For morphometric (quantitative) analysis, the following measurements were taken using ImageJ® (US National Institutes of Health, Bethesda, MD, USA) software at the level of osteonal units:

For the secondary osteons:

- Osteonal area;
- Osteonal perimeter;
- Vascular canal diameter and surface area.

The normality test (Kolmogorov–Smirnov) [63,64] did not show significant differences from the normal distribution in all illustrated series in the table below (Table 4).

Table 4. Metrical data for examined samples.

	H40 Area	H50 Area	Sn (Surgical Neck Area)
Secondary osteonal area (μm^2)	n = 40 SD = 4043.4 Mean: 18,414.2 Median: 17,702.3 *p*-value: 0.40478	n = 39 SD = 4264.4 Mean: 19,431.3 Median: 18,583.09 *p*-value: 0.63096	n = 39 SD = 5670.4 Mean: 23,196.99 Median: 21,688.61 *p*-value: 0.33228
Secondary osteon perimeter (μm)	n = 40 SD = 57.28 Mean: 496.9 Median: 492.11 *p*-value: 0.82778	n = 39 SD = 54.388 Mean: 505.366 Median: 505.048 *p*-value: 0.7777	n = 39 SD = 69.56 Mean: 549.37 Median: 536.15 *p*-value: 0.5498
Vascular canal area (μm^2)	n = 40 SD = 254.33 Mean: 674.96 Median: 616.11 *p*-value: 0.27436	n = 39 SD = 246.71 Mean: 639.60 Median: 561.1 *p*-value: 0.26501	n = 39 SD = 276.69 Mean: 628.02 Median: 598.16 *p*-value: 0.5035

4. Discussion

4.1. Gross Morphology and Osteometry

The general anatomy of the humerus in the *Chlorocebus sabaeus aethiopus* monkey from Saint Kitts and Nevis is similar to reported data in the classic anatomy books for mammals [16,20,65,66].

The proximal epiphysis in our subjects was characterised by the presence of the articular head of the humerus and the presence of two tubercles, the greater and smaller tubercle. The anatomical distribution of the structures and the general morphological aspects of the proximal end of the humerus allow us to consider that African green monkeys from Saint Kitts and Nevis have a terrestrial locomotor behaviour. Similar information has been reported for *Chlorocebus aethiopus* [20].

The general aspect of the articular head of the humerus can be an important indication of glenohumeral joint stability and mobility in primates and hominoids [9,67–70]. More than that, the aspect of this element can be used to differentiate between the primate species which use different types of locomotion.

In our studied specimens, the humeral head has been reported as an oval-shaped structure with low convexity and a flattened aspect, which articulates with the pear-shaped glenoid cavity of the scapula [71]. All of those aspects are similar to the reported data regarding monkeys using quadrupedal locomotion. Other scholars reported the same features of the head of the humerus in pig-tailed macaque (*Macaca nemestrina*), long-tailed macaque (*Macaca fascicularis*), and leaf monkeys. Remarkable differences have been reported between the anatomical aspects of the humeral head in the arboreal quadrupedal monkey and the terrestrial quadrupedal monkey. The first category is characterised by a slightly more rounded and pronounced articular head while the second category has similar aspects to those we described in our specimens [72]. These aspects confirm that our specimens usually are more prone to terrestrial quadrupedal locomotion.

Very different morphological aspects of the humeral head have been reported in gibbons, orangutans, gorillas, chimpanzees, and humans. The humeral head in all of these species was a globular and rounded structure. In monkeys, all of these characteristics are common for the suspensory type of locomotion [9,10,71,73]. In mangabeys and guenons, a hemispherical humeral head is described [20]. Even if they use only 11% of their time for suspensory locomotion, an intermediate aspect of the humeral head between the terrestrial and suspensory species has been reported in woolly monkeys as well [74].

Another morphological aspect which confirms that the African green monkey is a terrestrial quadrupedal species is the caudal orientation of the humeral head which fits with the cranio-caudal curvature of the glenoid cavity reported in Old World Monkeys and in African green monkeys [3,75].

The cranial continuation of the humeral head with the intertubercular groove, observed by us in *Chlorocebus sabaeus aethiopus* monkeys, is similar to the data reported in other terrestrial quadrupedal monkeys [10]. The macroscopic aspects of this structure allowed us to describe a U-shaped groove, located on the cranial aspect of the proximal epiphysis between the greater and lesser tubercles of the humerus. According to reported data, in baboons, chimpanzees, humans, and *Cebus libidinosus*, this space allows for the gliding of the tendon of origin of the lateral head of the biceps brachii muscle [76,77].

In the studied specimens, the greater and lesser tubercles occupied the medial and cranio-lateral surface of the proximal epiphysis. In some hominoids, migration of those two tubercles induces proximal torsion of the humerus. In great apes, this migration has a huge impact on the intertubercular groove, which migrates medially, and on the lesser tubercle which becomes very small [15,78].

The greater tubercle in macaques and proboscis monkeys is well developed, and different from African green monkeys, it extends above the head of the humerus [71].

In African green monkeys, the morphological aspects of the greater and lesser tubercles allowed for the identification of four insertion points for the rotator cuff muscles: three insertion points at the level of the greater tubercle, for the supraspinatus, infraspinatus, and teres minor muscles, and the last one on the lesser tubercle for the subscapularis muscle. Similar aspects have been reported in humans, baboons, chimpanzees, mangabeys, guenons, and Cebus [20,79].

According to [9], the locomotor behaviour can influence the anatomo-topographical distribution of the insertion points of the rotator cuff muscles on the greater tubercle, and differences between fully quadrupedal and fully suspensory primates can be identified.

The lateral disposition of the insertion point of the infraspinatus muscle, reported by us in *Chlorocebus sabaeus aethiopus*, confirms the reported data in cercopithecoids [80].

Similar morphological aspects of the greater tubercle reported by us in the present study have been reported in *Cercocebus galericus*, *Cercopithecus aethiops*, and other mangabey monkeys [20]. The same author has reported an incomplete greater tubercle in Cercocebus albigena, where the caudo-lateral segment lacks.

In humans, it seems that the reduced size of the rotator cuff muscles is in direct correlation with the reduced size of the greater tubercle [9,15] and that the supraspinatus muscle is very reduced in size [80]. In *Cebus libidinosus*, other muscles such as *m.pectoralis abdominis* and *m.pectoralis minor*, which are not rotator cuff muscles, have as insertion points the greater tubercle of the humerus [76].

The lesser tubercle of the humerus in the studied species was well developed, and according to the reported data, it serves as the insertion point for the subscapularis muscle [78]. In mangabeys and guenons, the literature has reported that the macroscopic and morphometric aspects of the lesser tubercle were more constant compared to those of the greater tubercle. A well-developed lesser tubercle was reported in baboons, and it is strong proof of the significant role of the subscapularis muscle in quadrupedal terrestrial locomotion [20]. Similar features were described in chimpanzees and humans [79].

In our study, we noted that the humeral diaphysis showed an obvious craniolateral curvature in the proximal third. Also, the crests and tuberosities were very well developed in *Chlorocebus sabaeus aethiopus* monkeys. Similar features have been observed in both pig-tailed and long-tailed macaques and in leaf monkeys [71].

Previous studies have reported differences in humeral curvature between terrestrial and arboreal quadrupedal species of primates. In terrestrial species, the action of the long head of the triceps muscle (*M. triceps brachii—Caput longum*) together with the teres major muscle (*M. teres major*), the scapular part of the deltoid muscle (*M. deltoideus—pars scapularis*), and the latissimus dorsi (*M. latissimus dorsi*) influence the cranio-lateral curvature of the humeral shaft [81,82].

For comparison, the only comparable data (as mentioned in Table 1 above) we have at our disposal are those related to the angulation remeasures for humans (based on the specimens in the Museum of Anatomy of Faculty of Veterinary Medicine Cluj-Napoca, Romania) and some online available published specimens of different monkey species. For *Homo sapiens*, a value of 175 and 170 degrees was recorded; for orangutan (*Pogo pygmaeus*), a comparative value of 165 degrees was noted, and as for chimpanzees (*Pan troglodites*), an angulation of 178 and 173 degrees was noted. These are values for the Hominidae group. As for a member of Cercopithhidae, *Macacus rhesus*, we managed to remeasure a value of 153 degrees, the closest value to the recorded angle in *Chlorocebus sabeus* (154–155 degrees) (Table 1).

In arboreal species, the humeral shaft is double-curved. Contraction of the scapular part of the deltoid muscle, teres major, and latissimus dorsi muscles causes the cranio-lateral curvature of the proximal third of the humeral diaphysis. The action of muscles involved in climbing, clinging, and grasping behaviours (flexor and extensor muscles of the carpal and digital regions and brachioradialis muscle (*M. brachioradialis*)) induces distal cranio-caudal curvature of the humeral body. Similar data have been reported in gibbons [65,71,81,83,84].

Our investigated specimens showed an evident surgical neck with a very well-developed bony crest and tuberosities for muscular insertion, like for cercopithecines [20]. Reduced muscular crests and tuberosities were reported in orangutans [71] and in the common marmoset [85]. According to [76], in *Cebus libidinosus*, the muscular tuberosities and crests, located on the humeral shaft, serve as insertion points for the deltoideus, teres major, latissimus dorsi, coracobrachialis (*M. coracobrachialis*), and *pectoralis major* muscles, and provide origin points for the brachialis muscle (*M. brachialis*) and the lateral (*M. triceps brachii—caput laterale*) and medial head (*M. triceps brachii—caput mediale*) of the triceps muscle.

The different types of locomotion have a huge impact at the level of the distal epiphysis of the humerus [9,86,87]. The humeral condyle in African green monkeys is a component of two sub-joints: humeroulnar (*Articulatio humeroulnaris*) and humeroradial (*Articulatio humeroradialis*) joints. Because of a reduced medial trochlear keel and a very small zona conoidea, the demarcation between the humeral trochlea and the humeral capitulum is inconspicuous. The presence of a deep waist between the zona conoidea and the medial trochlear keel has been reported in orangutans [71]. Our results confirm the data reported by [86] which sustain that in Old World monkeys, the stabilization features of the humeroradial joint are absent.

As we reported in our study, the medial epicondyle was well developed and showed medial caudal retroflexion. These features are specific for terrestrial monkeys such as the Papio genus [88]. The development of the epicondyle is directly related to the flexor muscle of the carpus and digit development and activity [20]. Opposite to our findings, a large and medially directed medial epicondyle is typical for arboreal species [89,90]. Supratrochlear communication, reported between the olecranon fossa and radial fossa in a few specimens of *Chlorocebus sabaeus aethiopus*, is lacking in common marmoset [90]. The existence of the supratrochlear foramen along with a deep olecranon fossa increases the range of flexion [91].

Authors should discuss the results and how they can be interpreted from the perspective of previous studies and the working hypotheses. The findings and their implications should be discussed in the broadest context possible. Future research directions may also be highlighted.

4.2. *Micromorphology and Histomorphometric Data Interpretation*

The examined materials present a typical histological arrangement for primates, with the prevalence of lamellar bone arrangement. This blend of primary avascular and vascular primary bone (the latter prevailing though) was listed as a general feature for non-human primates [32]. This is an indicator of a slow(er) bone deposition process, which is quite typical for this group [23,54].

The secondary bone type that prevails in the structure is the dense osteonal (Haversian) one, with osteons showing no or little organization, almost like being randomly distributed through the cortex. The scattered osteonal structures also have quite a constant presence, with a smaller ratio when compared to the previous way of organization. This feature is also noted as a non-human characteristic among primates' bone structure [32]. Although attempts to use the density of the osteonal system as a differential feature for different locomotive patterns (arboreal, terrestrial or suspensory) were mentioned in the literature [33], evaluating the number of Haversian systems per square millimetres yielded such diverse figures in the studied samples that we decided not to attempt such a comparison.

Drifting osteons (DOs) are typically absent in most mammalian species. Their existence was noted in several human-origin bone specimens [62], being initially considered a discriminating factor in identifying the human bone histologically. The explanation given by literature sources for this type of peculiar osteonal pattern points to the limited vascular pathway and the bone remodelling units that force the microcrack repair in different directions, allowing for movement (or transverse drifting) due to vascular plasticity and vascular connectivity between osteonal units [60,62]. This arrangement has also been noted in baboons and other primates [62].

Osteon banding was earlier noted in other mammals and non-human primates [22,32,34–36] and it seems to display certain, very fine differential features among human and mammalian bone [92]. This type of arrangement is associated with rapid growth (seen in subadults), and it seems to disappear during life [32].

A plot of the values recorded for the secondary osteon size and the vascular canal area, differentiated for the different areas of choice of the bone, shows the existence of some large osteonal units in the SN area only, as the values for H40 and H50 have a great degree of overlapping (Figure 10). This situation was also noticed when the raw data were processed, mainly for the case of the average secondary osteonal area.

Figure 10. A plot of the osteonal area vs. the vascular canal area based on their region of origin (H40, H50, SN).

The existence of giant osteons, according to literature sources, might be linked with the smaller loads and reduced tension environment, showing adaptation for resisting deleterious shear stress [93]. This specific structural element might also be associated with the pathology expressed at the surgical neck area [94] or linked with remodelling phenomena, having no association with age and gender [95].

To check whether there are significant statistical differences among the values recorded for the osteonal area and the vascular canal, a more complex set of tests was used for the difference of means—the one-way ANOVA and Tukey, Schefee, Bonferroni, and Holm tests. The results (with limitations given by the relatively small samples) for the investigated sets are listed in Table 5.

As noted, the differences expressed at the level of the secondary osteonal area are not notable statistically in the case of the recorded values for the vascular canal area. This variation in terms of the secondary osteonal area and diameters is supposedly related to strain, due to a mechanism still unknown [96,97]. The average values for secondary osteonal units are still of importance, being allegedly used as a basic reference for species-linked data, illustrating differential aspects among species more or less related, mainly in conjunction with qualitative data.

Table 5. Statistical tests performed on the micrometrical collected data (Ar = area).

	Anova	Tukey HSD Results	Scheffe Multiple Comparison	Bonferroni and Holm
Secondary osteonal area	The p-value corresponding to the F-statistic is 0.05, suggesting that one or more treatments are significantly different.	H40 Ar vs. H50 Ar—inisgnificant H40 Ar vs. SN Ar—**, $p < 0.01$ H50 Ar vs. SN Ar—**, $p < 0.01$	H40 Ar vs. H50 Ar—insignificant H40 Ar vs. SN Ar—**, $p < 0.01$ H50 Ar vs. SN AR—*, $p < 0.05$	H40 Ar vs. H50 Ar—insignificant H40 Ar vs. SN Ar—**, $p < 0.01$ H50 Ar vs. SN AR—**, $p < 0.01$
Vascular canal area	The p-value corresponding to the F-statistic of the one-way ANOVA is higher than 0.05, suggesting that the treatments are not significantly different for that level of significance.			

5. Conclusions

Our study provides morphologic, morphometric, and structural features regarding the humerus in *Chlorocebus sabaeus aethiopus* from the Saint Kitts and Nevis Federation.

All of the morphological aspects of the humerus confirm the terrestrial locomotor behaviour in the studied specimens and are similar to other data reported for other Old World Monkey species.

The osteometrical data complete the series of metrical data for different species, offering a base for other comparative approaches.

The micromorphological data provide a series of insights into this little studied primate, such as the similitude with other primates in terms of histological arrangements and the exact histomorphometric (or specific) elements in terms of the secondary periostal bone structures (osteons).

Some other interesting details are highlighted, such as the existence of slightly different quotas of the typical arrangements of bone types related to the regions of bones. The mentioning of the structures called super-osteons and drifting osteons makes nothing but interesting the potential of morphological study of this species and the correlations and applicability of different studies on this species, with impacts on the different areas of the medical and biological sciences.

Author Contributions: Conceptualization, C.O.M., A.I.G. and P.B.; methodology, C.O.M., A.I.G. and P.B.; software, A.I.G. and C.L.; validation, C.O.M., A.I.G., W.B.L., P.B., F.G.S. and C.C.D.; formal analysis, C.O.M., A.I.G., W.B.L., C.C.D. and P.B.; investigation, C.O.M., A.I.G., F.G.S., P.B. and C.C.D.; resources, C.O.M., A.I.G. and P.B.; data curation, A.I.G.; writing—original draft preparation, C.O.M., A.I.G. and W.B.L.; writing—review and editing, C.O.M., A.I.G., W.B.L., F.G.S. and P.B.; visualization, C.O.M., A.I.G., W.B.L., F.G.S., P.B., C.L. and C.C.D.; supervision, C.O.M., A.I.G. and P.B.; project administration, C.O.M. and A.I.G.; funding acquisition, C.O.M. and A.I.G. All authors have read and agreed to the published version of the manuscript.

Funding: This project received no specific funding and falls under RUSVM's Center for Integrative Mammalian Research.

Institutional Review Board Statement: The animal study protocol was approved by the Institutional Review Board of the Ross University Institutional Animal Care and Use Committee (TSU10.27.2023 CM) following the IACUC regulations.

Informed Consent Statement: Not applicable.

Data Availability Statement: The data that support the findings of this study are available on reasonable request from the corresponding author, G.A.I.

Acknowledgments: The authors would like to express their thanks for the support given by Roger Hancock, the curator of the biological collection in Saint Kitts and to Hilchie David, manager of the Histology Lab, Ross University, School of Veterinary Medicine, Saint Kitts and Nevis.

Conflicts of Interest: The authors declare no conflicts of interest.

References

1. Schoch, C.L.; Ciufo, S.; Domrachev, M.; Hotton, C.L.; Kannan, S.; Khovanskaya, R.; Leipe, D.; McVeigh, R.; O'Neill, K.; Robbertse, B.; et al. NCBI Taxonomy: A Comprehensive Update on Curation, Resources and Tools. *Database* **2020**, *2020*, baaa062. [CrossRef] [PubMed]
2. Young, A.N.; du Plessis, W.M.; Rodriguez, D.; Beierschmitt, A. Thoracic Radiographic Anatomy in Vervet Monkeys (*Chlorocebus sabaeus*). *J. Med. Primatol.* **2013**, *42*, 310–317. [CrossRef] [PubMed]
3. Martonos, C.O.; Little, W.B.; Lațiu, C.; Gudea, A.I.; Stan, F.G.; Dezdrobitu, C.C. Morphological and Morphometrical Features of the Scapula in *Chlorocebus sabaeus* from Saint Kitts and Nevis Islands. *Rev. Romana Med. Vet.* **2024**, *34/1*, 85–94.
4. Martonos, C.O.; Gudea, A.I.; Ratiu, I.A.; Stan, F.G.; Bolfă, P.; Little, W.B.; Dezdrobitu, C.C. Anatomical, Histological, and Morphometrical Investigations of the Auditory Ossicles in *Chlorocebus aethiops* Sabaeus from Saint Kitts Island. *Biology* **2023**, *12*, 631. [CrossRef]
5. Freimer, N.B.; Service, S.K.; Ophoff, R.A.; Jasinska, A.J.; McKee, K.; Villeneuve, A.; Belisle, A.; Bailey, J.N.; Breidenthal, S.E.; Jorgensen, M.J.; et al. A Quantitative Trait Locus for Variation in Dopamine Metabolism Mapped in a Primate Model Using Reference Sequences from Related Species. *Proc. Natl. Acad. Sci. USA* **2007**, *104*, 15811–15816. [CrossRef] [PubMed]
6. Branco, É.; Silva, K.S.M.; De Fátima Assis, M.; De Oliveira, E.H.C.; De Lima, A.R.; De Aquino, A.C.G.; Cabral, R.; Miglino, M.A. Protocol for Collection and Separation of Bone Marrow Mononuclear Cells in *Chlorocebus aethiops*. *Pesqui. Vet. Bras.* **2016**, *36*, 119–122. [CrossRef]
7. Paschalis, E.P.; Gamsjaeger, S.; Burr, D.B. Bone Quality in an Ovariectomized Monkey Animal Model Treated with Two Doses of Teriparatide for Either 18 Months, or 12 Months Followed by Withdrawal for 6 Months. *Bone* **2022**, *158*, 116366. [CrossRef]
8. Bonucci, E.; Ballanti, P. Osteoporosis-Bone Remodeling and Animal Models. *Toxicol. Pathol.* **2014**, *42*, 957–969. [CrossRef]
9. Arias-Martorell, J.; Tallman, M.; Potau, J.M.; Bello-Hellegouarch, G.; Pérez-Pérez, A. Shape Analysis of the Proximal Humerus in Orthograde and Semi-Orthograde Primates: Correlates of Suspensory Behavior. *Am. J. Primatol.* **2015**, *77*, 1–19. [CrossRef]
10. Arias-Martorell, J. The Morphology and Evolutionary History of the Glenohumeral Joint of Hominoids: A Review. *Ecol. Evol.* **2019**, *9*, 703–722. [CrossRef]
11. Costes, P.; Klein, E.; Delapré, A.; Houssin, C.; Nicolas, V.; Cornette, R. Comparative Morpho-Functional Analysis of the Humerus and Ulna in Three Western European Moles Species of the Genus Talpa, Including the Newly Described *T. aquitania*. *J. Anat.* **2023**, *242*, 257–276. [CrossRef] [PubMed]
12. Biewener, A.A.; Patek, S.N. *Animal Locomotion*, 2nd ed.; Oxford University Press: Oxford, UK, 2018; ISBN 9780198743156.
13. Martín-Serra, A.; Figueirido, B.; Palmqvist, P. A Three-Dimensional Analysis of the Morphological Evolution and Locomotor Behaviour of the Carnivoran Hind Limb. *BMC Evol. Biol.* **2014**, *14*, 129. [CrossRef]
14. Polly, P.D. Limbs in Mammalian Evolutio. In *Fins into Limbs: Evolution, Development, and Transformation*; Brian, K.H., Ed.; University of Chicago Press: Chicago, IL, USA, 2007; p. 258.
15. Arias-Martorell, J.; Potau, J.M.; Bello-Hellegouarch, G.; Pastor, J.F.; Pérez-Pérez, A. 3D Geometric Morphometric Analysis of the Proximal Epiphysis of the Hominoid Humerus. *J. Anat.* **2012**, *221*, 394–405. [CrossRef] [PubMed]
16. Evans, H.E.; de Lahunta, A. *Miller's Anatomy of the Dog*; W.B. Saunders Company: Philadelphia, PA, USA; London, UK; Toronto, ON, Canada; Montreal, QC, Canada; Sydney, Australia; Tokyo, Japan, 2012.
17. Johnston, R. *Capitular Morphology in Primates and Fossils: 3-D Measurements of the Capitulum and Possible Implications for Reconstructing Locomotor Adaptations in Fossil Primates*; Columbia Graduate School, University of Missouri: Columbia, MO, USA, 2014.
18. Cubo, J. Pattern and Process in Constructional Morphology. *Evol. Dev.* **2004**, *6*, 131–133. [CrossRef] [PubMed]
19. Henderson, K.; Pantinople, J.; McCabe, K.; Richards, H.L.; Milne, N. Forelimb Bone Curvature in Terrestrial and Arboreal Mammals. *PeerJ* **2017**, *5*, e3229. [CrossRef]
20. Nakatsukasa, M. Morphology of the Humerus and Femur in African Mangabeys and Guenons: Functional Adaptation and Implications for the Evolution of Positional Behavior. *Afr. Study Monogr.* **1994**, *21*, 1–61. [CrossRef]
21. Enlow, D.H.; Brown, S.O. A Comparative Histological Study of Fossil and Recent Bone Tissues. Part II. *Tex. J. Sci.* **1956**, *9*, 186–214.
22. Cuijpers, A.G.F.M. Species Identification in Archaeology. Ph.D. Thesis, Faculty of Earth and Life Sciences, University of Amsterdam, Amsterdam, The Netherlands, 2009; p. aprox 150.
23. Cuijpers, A.G.F.M. Distinguishing between the Bone Fragments of Medium-Sized Mammals and Children. A Histological Identification Method for Archaeology. *Antropol. Anz.* **2009**, *67*, 181–203. [CrossRef]
24. Cuijpers, A.G.F.M. Histological Identification of Bone Fragments in Archaeology: Telling Humans Apart from Horses and Cattle. *Int. J. Osteoarchaeol.* **2006**, *16*, 456–480. [CrossRef]
25. Martiniakova, M.; Grosskopf, B.; Omelka, R.; Vondrakova, M.; Bauerova, M. Differences among Species in Compact Bone Tissue Microstructure of Mammalian Skeleton: Use of a Discriminat Function Analysis for Species Identification. *J. Forensic Sci.* **2006**, *51*, 1235–1239. [CrossRef]
26. Martiniakova, M.; Grosskopf, B.; Vondrakova, M.; Omelka, R.; Fabis, M. Differences in Femoral Compact Bone Tissue Microscopic Structure between Adult Cows (*Bos taurus*) and Pigs (*Sus scrofa domestica*). *Anat. Histol. Embryol.* **2006**, *35*, 167–170. [CrossRef] [PubMed]
27. Martiniakova, M.; Grosskopf, B.; Omelka, R.; Vondrakova, M.; Bauerova, M. Histological Analysis of Ovine Compact Bone Tissue. *J. Vet. Med. Sci.* **2007**, *69*, 409–411. [CrossRef] [PubMed]
28. Rajtova, V.; Briancin, J.; Kokardova, M. Lamellar Bone Structure in Small Ruminats. *Folia Vet.* **1995**, *3–4*, 123–126.

29. Jowsey, J. Studies of Haversian Systems in Man and Some Animals. *J. Anat.* **1966**, *100*, 857–864.
30. Dittmann, K. Histomorphometrische Untersuchung Der Knochen Mikrostruktur von Primaten Und Haustieren Mit Them Ziel Der Speziesidentifikaton Unter Berucksichtigung von Domestikationeffecten. *Anthrop. Anz.* **2003**, *61*, 175–188. [CrossRef]
31. Dittmann, K.; Grupe, G.; Manhart, H.; Peters, J.; Strott, N. Histomorphometry of Mammalian and Avian Compact Bone. Ph.D. Thesis, Facultat fur Biologie, Bereich Biodiversitatsforschung, Martinsreid, Germany, 2008; pp. 49–101.
32. Brits, D.; Steyn, M.; L'Abbé, E.N. A Histomorphological Analysis of Human and Non-Human Femora. *Int. J. Legal Med.* **2014**, *128*, 369–377. [CrossRef]
33. Hillier, M.L.; Lynne, S.B. Differentiating Human Bone from Animal Bone: A Review of Histological Methods. *J. Forensic Sci.* **2007**, *52*, 249–263. [CrossRef]
34. Gudea, A.I.; Stefan, A.C. Histomorphometric, Fractal and Lacunarity Comparative Analysis of Sheep (*Ovis aries*), Goat (*Capra hircus*) and Roe Deer (*Capreolus capreolus*) Compact Bone Samples. *Folia Morphol.* **2013**, *72*, 239–248. [CrossRef]
35. Gudea, A.; Stefan, A. Histomorphometric and Fractal Analysis of Femoral, Tibial, and Metatarsal Compact Bone Samples in Sheep (*Ovis aries*), Goat (*Capra hircus*), and Roe Deer (*Capreolus capreolus*). *Turk. J. Vet. Anim. Sci.* **2015**, *39*, 528–536. [CrossRef]
36. Zedda, M.; Palombo, M.R.; Brits, D.; Carcupino, M.; Sathé, V.; Cacchioli, A.; Farina, V. Differences in Femoral Morphology between Sheep (*Ovis aries*) and Goat (*Capra hircus*): Macroscopic and Microscopic Observations. *Zoomorphology* **2017**, *136*, 145–158. [CrossRef]
37. Zedda, M.; Lepore, G.; Biggio, G.P.; Gadau, S.; Mura, E.; Farina, V. Morphology, Morphometry and Spatial Distribution of Secondary Osteons in Equine Femur. *J. Vet. Med. Ser. C Anat. Histol. Embryol.* **2015**, *44*, 328–332. [CrossRef] [PubMed]
38. Zedda, M. The Arrangement of the Osteons and Kepler's Conjecture. *Appl. Sci.* **2023**, *13*, 5170. [CrossRef]
39. Cvetkovic, V.J.; Najman, S.J.; Rajkovic, J.S.; Zabar, A.L.J.; Vasiljevic, P.J.; Djordjevic, L.J.B.; Trajanovic, M.D. A Comparison of the Microarchitecture of Lower Limb Long Bones between Some Animal Models and Humans: A Review. *Veterinární Med.* **2013**, *58*, 339–351. [CrossRef]
40. Lasota-Moskalewska, A.; Moskalewski, S. Microscopic Comparrison of Bones from Medieval Domestic and Wild Pigs. *Ossa* **1982**, *7*, 173–178.
41. Dammers, K. Using Osteohistology for Ageing and Sexing. In Proceedings of the 9-th ICAZ Conference-Recent, Advances in Ageing and Sexing Animal Bones 2002, Online, 23 September 2021; pp. 9–40.
42. Cattaneo, C.; Porta, D.; Gibelli, D.; Gambe, C. Histological Determination of the Human Origin of Bone Fragments. *J. Forensic Sci.* **2009**, *54*, 531–533. [CrossRef]
43. Felder, A.A.; Phillips, C.; Cornish, H.; Cooke, M.; Hutchinson, J.R.; Doube, M. Secondary Osteons Scale Allometrically in Mammalian Humerus and Femur. *R. Soc. Open Sci.* **2017**, *4*, 170431. [CrossRef]
44. Yu, W.L.; Cen, H.P.; Wu, X.G.; Guo, Y.; Li, C.X.; Wang, Y.Q.; Chen, W.Y. Finite Element Study of the Effect of Osteon Morphology Variation Related Ageing, Osteoporosis, or Physical Activity Level on Its Poroelastic Behaviors. *J. Hard Tissue Biol.* **2018**, *27*, 333–342. [CrossRef]
45. Bernhard, A.; Milovanovic, P.; Zimmermann, E.A.; Hahn, M.; Djonic, D.; Krause, M.; Breer, S.; Püschel, K.; Djuric, M.; Amling, M.; et al. Micro-Morphological Properties of Osteons Reveal Changes in Cortical Bone Stability during Aging, Osteoporosis, and Bisphosphonate Treatment in Women. *Osteoporos. Int.* **2013**, *24*, 2671–2680. [CrossRef]
46. Kim, J.N.; Lee, J.Y.; Shin, K.J.; Gil, Y.C.; Koh, K.S.; Song, W.C. Haversian System of Compact Bone and Comparison between Endosteal and Periosteal Sides Using Three-Dimensional Reconstruction in Rat. *Anat. Cell Biol.* **2015**, *48*, 258. [CrossRef]
47. Rueden, C.T.; Schindelin, J.; Hiner, M.C.; DeZonia, B.E.; Walter, A.E.; Arena, E.T.; Eliceiri, K.W. ImageJ2: ImageJ for the next generation of scientific image data. *BMC Bioinform.* **2017**, *18*, 529. [CrossRef]
48. The GIMP Development Team. GIMP 2.10.38. Retreived August 2024. Available online: https://www.gimp.org/ (accessed on 26 August 2024).
49. Waibl, H.; Gasse, H.; Constantinescu, G.; Hashimoto, Y.; Simoens, P. *Nomina Anatomica Veterinaria*, 6th ed.; International Committee, Veterinary Gross Anatomical Nomenclature: Hanover, Germany; Ghent, Belgium; Columbia, MO, USA; Rio de Janeiro, Brazil, 2017.
50. Ogedengbe, O.O.; Ajayi, S.A.; Komolafe, O.A.; Zaw, A.K.; Naidu, E.C.S.; Azu, O.O. Sex Determination Using Humeral Dimensions in a Sample from KwaZulu-Natal: An Osteometric Study. *Anat. Cell Biol.* **2017**, *50*, 180–186. [CrossRef] [PubMed]
51. Akman, Ş.D.; Karakaş, P.; Bozkir, M.G. The Morphometric Measurements of Humerus Segments. *Turk. J. Med. Sci.* **2006**, *36*, 81–85.
52. Patel, B.A.; Ruff, C.B.; Simons, E.L.R.; Organ, J.M. Humeral Cross-Sectional Shape in Suspensory Primates and Sloths. *Anat. Rec.* **2013**, *296*, 545–556. [CrossRef] [PubMed]
53. McManus, J.F.A.; Mowry, R.W. *Staining Methods: Histologic and Histochemical*; Hoeber: New York, NY, USA, 1960.
54. Cuijpers, A.G.F.M.; Lauwerier, R.C.G.M. Differentiating between Bone Fragments from Horses and Cattle: A Histological Identification Method for Archaeology. *Environ. Archaeol.* **2008**, *13*, 163–173. [CrossRef]
55. Morris, Z.H. Quantitative and Spatial Analysis of the Microscopic Bone Structures Od Deer (*Odocoileus virginianus*) Dog (*Canis familiaris*) and Pig (*Sus scrofa domestica*). Master's Thesis, University of Toronto, Toronto, QC, Canada, 2007.
56. Cooke, K.; Mahoney, P.; Miszkiewicz, J.J. Secondary osteon variants and remodeling in human bone. *Anat. Rec.* **2021**, *305*, 1229–1313. [CrossRef]
57. Rose, M.D. Another Look at the Anthropoid Elbow. *J. Hum. Evol.* **1988**, *17*, 193–224. [CrossRef]

58. French, K.M.C.; Crowder, C.; Crabtree, P.J. Integrating Histology in the Analysis of Multispecies Cremations: A Case Study from Early Medieval England. *Int. J. Osteoarchaeol.* **2022**, *32*, 1253–1263. [CrossRef]
59. Cooke, K.M.; Mahoney, P.; Miszkiewicz, J.J. Secondary Osteon Variants and Remodeling in Human Bone. *Anat. Rec.* **2022**, *305*, 1299–1315. [CrossRef]
60. Robling, A.G.; Stout, S.D. Morphology of the Drifting Osteon. *Cells Tissues Organs* **1999**, *164*, 192–204. [CrossRef]
61. Andronowski, J.M.; Pratt, I.V.; Cooper, D.M.L. Occurrence of Osteon Banding in Adult Human Cortical Bone. *Am. J. Phys. Anthropol.* **2017**, *164*, 635–642. [CrossRef]
62. French, K.M.; Mavroudas, S.R.; Dominguez, V.M. Prevalence of Drifting Osteons Distinguishes Human Bone. *PLoS ONE* **2024**, *19*, e0298029. [CrossRef] [PubMed]
63. Statistics Online—Checks Assumptions, Interprets Results. Available online: https://www.statskingdom.com/index.html (accessed on 9 January 2024).
64. Quick Statistics Calculators. Available online: https://www.socscistatistics.com/tests/ (accessed on 2 December 2023).
65. Evans, H.E.; de Lahunta, A. *Guide to the Dissection of the Dog*; Saunders Elsevier: St. Louis, MO, USA, 2010.
66. Dyce, K.M.; Sack, O.W.; Wensig, C.J.G. *Textbook of Veterinary Anatom*, 4th ed.; Saunders Elsevier: St Louis, MO, USA, 2010; ISBN 978-1-4160-6607-1.
67. Rose, M.D. New Postcranial Specimens of Catarrhines from the Middle Miocene Chinji Formation, Pakistan: Descriptions and a Discussion of Proximal Humeral Functional Morphology in Anthropoids. *J. Hum. Evol.* **1989**, *18*, 131–162. [CrossRef]
68. Rafferty, K.L.; Ruff, C.B. Articular Structure and Function in Hylobates, Colobus, and Papio. *Am. J. Phys. Anthropol.* **1994**, *94*, 395–408. [CrossRef] [PubMed]
69. Ruff, C.B.; Runestad, J.A. Primate Limb Bone Structural Adaptations. *Annu. Rev. Anthropol.* **1992**, *21*, 407–433. [CrossRef]
70. Lap, K.C. The Range of Passive Arm Circumduction in Primates: Do Hominoids Really Have More Mobile Shoulders? *Am. J. Phys. Anthropol.* **2008**, *136*, 265–277. [CrossRef]
71. Pipper, K.K.-K.; John, P. Identification of Morphological Variation in the Humeri of Bornean Primates and Its Application to Zooarchaeology. *Archaeofauna* **2004**, *13*, 85–95.
72. Preuschoft, H.; Hohn, B.; Scherf, H.; Schmidt, M.; Krause, C.; Witzel, U. Functional Analysis of the Primate Shoulder. *Int. J. Primatol.* **2010**, *31*, 301–320. [CrossRef]
73. Jungers, W.L. Relative Joint Size and Hominoid Locomotor Adaptations with Implications for the Evolution of Hominid Bipedalism. *J. Hum. Evol.* **1988**, *17*, 247–265. [CrossRef]
74. Kagaya, M. Glenohumeral Joint Surface Characters and Its Relation to Forelimb Suspensory Behavior in Three Ateline Primates, Ateles, Lagothrix, and Alouatta. *Anthropol. Sci.* **2007**, *115*, 17–23. [CrossRef]
75. MaClatchy, L.; Gebo, D.; Kityo, R.; Pilbeam, D. Postcranial Functional Morphology of Morotopithecus Bishopi, with Implications for the Evolution of Modern Ape Locomotion. *J. Hum. Evol.* **2000**, *39*, 159–183. [CrossRef]
76. Aversi Ferreira, T.A.; Pereira de Paula, J.; do Prado, Y.C.L.; Lima e Silva, M.S.; da Mata, R. Anatomy of the Shouder and Arm Muscles of Cebus Libidinosus. *Braz. J. Morphol. Sci.* **2007**, 63–74.
77. Moore, K.L.; Dalley, A.F.; Agur, A.M. *Moore-Anatomia Orientada Para a Clínica*; Woltes, K., Ed.; Guanabara Koogan: Rio de Janeiro, Brazil, 2014; ISBN 9788578110796.
78. Aiello, L.; Dean, C. *An Introduction to Human Evolutionary Anatomy*; Academic Press Inc.: Cambridge, MA, USA, 1990.
79. Aversi-Ferreira, T.A.; Pererira-de-Paula, J.; Lima-e-Silva, M.D.S.; Silva, Z. Anatomia Das Artérias Do Braço Do Macaco Cebus Libidinosus (Rylands et al., 2000). *Acta Sci. Biol. Sci.* **2007**, *29*, 247–254. [CrossRef]
80. Larson, S.G. New Characters for the Functional Interpretation of Primate Scapulae and Proximal Humeri. *Am. J. Phys. Anthropol.* **1995**, *98*, 13–35. [CrossRef] [PubMed]
81. Milne, N.; Granatosky, M.C. Ulna Curvature in Arboreal and Terrestrial Primates. *J. Mamm. Evol.* **2021**, *28*, 897–909. [CrossRef]
82. Harrison, S.M.; Whitton, R.C.; King, M.; Haussler, K.K.; Kawcak, C.E.; Stover, S.M.; Pandy, M.G. Forelimb Muscle Activity during Equine Locomotion. *J. Exp. Biol.* **2012**, *215*, 2980–2991. [CrossRef]
83. Youlatos, D. Functional Anatomy of Forelimb Muscles in Guianan Atelines (Platyrrhini: Primates). *Ann. Des Sci. Nat.—Zool. Et Biol. Anim.* **2000**, *21*, 137–151. [CrossRef]
84. Taylor, M.E. The Functional Anatomy of the Forelimb of Some African Viverridae (Carnivora). *J. Morphol.* **1974**, *143*, 307–335. [CrossRef]
85. Casteleyn, C.; Bakker, J. The Anatomy of the Common Marmoset. In *The Common Marmoset in Captivity and Biomedical Research*; Academic Press: Cambridge, MA, USA, 2019. [CrossRef]
86. Zelazny, K.G. *Morphological Variation in the Distal Humerus of Extant Hominids and Fossil Hominins*; Johns Hopkins University: Baltimore, MD, USA, 2019.
87. Gebo, D.L. Climbing, Brachiation, and Terrestrial Quadrupedalism: Historical Precursors of Hominid Bipedalism. *Am. J. Phys. Anthropol.* **1996**, *101*, 55–92. [CrossRef]
88. Vagtborg, H. *The Baboon in Medical Research, II*; University of Texas Press: Austin, TX, USA, 1967.
89. Fleagle, J. *Primate Adaptation and Evolution*; Academic Press Inc.: Cambridge, MA, USA, 2013.
90. Casteleyn, C.; Bakker, J.; Breugelmans, S.; Kondova, I.; Saunders, J.; Langermans, J.A.M.; Cornillie, P.; Van den Broeck, W.; Van Loo, D.; Van Hoorebeke, L.; et al. Anatomical Description and Morphometry of the Skeleton of the Common Marmoset (*Callithrix jacchus*). *Lab. Anim.* **2012**, *46*, 152–160. [CrossRef]

91. Rose, M.D. Hominoid postcranial specimens from the Middle Miocene Chinji Formation, Pakistan. *J. Hum. Evol.* **1984**, *13*, 503–516. [CrossRef]
92. Mulhern, D.M.; Ubelaker, D.H. Differences in Osteon Banding between Human and Nonhuman Bone. *J. Forensic Sci.* **2015**, *46*, 220–222. [CrossRef]
93. Skedros, J.G.; Keenan, K.E.; Williams, T.J.; Kiser, C.J. Secondary Osteon Size and Collagen/Lamellar Organization ("Osteon Morphotypes") Are Not Coupled, but Potentially Adapt Independently for Local Strain Mode or Magnitude. *J. Struct. Biol.* **2013**, *181*, 95–107. [CrossRef] [PubMed]
94. Barakat, A.; Hatrick, N.C. Case Report of a Surgical Neck Humerus Fracture Due to Strength Training. *Trauma. Case Rep.* **2019**, *21*, 100205. [CrossRef] [PubMed]
95. Bell, K.L.; Loveridge, N.; Reeve, J.; Thomas, C.D.L.; Feik, S.A.; Clement, J.G. Super-Osteons (Remodeling Clusters) in the Cortex of the Femoral Shaft: Influence of Age and Gender. *Anat. Rec.* **2001**, *264*, 378–386. [CrossRef]
96. van Oers, R.F.M.; Ruimerman, R.; van Rietbergen, B.; Hilbers, P.A.J.; Huiskes, R. Relating Osteon Diameter to Strain. *Bone* **2008**, *43*, 476–482. [CrossRef]
97. Skedros, J.G.; Clark, G.C.; Sorenson, S.M.; Taylor, K.W.; Qiu, S. Analysis of the Effect of Osteon Diameter on the Potential Relationship of Osteocyte Lacuna Density and Osteon Wall Thickness. *Anat. Rec.* **2011**, *294*, 1472–1485. [CrossRef]

life

MDPI

Article

Correlation Between Hyperlipidemia-Related Diseases and Thorax/Thigh Circumference Ratio Along with Body Condition Score in Dogs Focusing on Molecular Mechanism: A Pilot Study and Literature Review

Kyuhyung Choi [1,2]

[1] Department of Veterinary Pathology, College of Veterinary Medicine, Seoul National University, Seoul 08826, Republic of Korea; cokonut@naver.com or kyudac@snu.ac.kr

[2] Bundang New York Animal Hospital, Seongnam 13637, Republic of Korea

Abstract: There are some limitations to using the body condition score (BCS) for client education to prevent obesity, hyperlipidemia-related diseases, and orthopedic diseases in dogs because it is hard to quantify in detail. Especially in small dogs, patellar luxation is a common orthopedic disease that is related to obesity and the hind leg muscle. In this pilot study, the author evaluated the thorax/thigh circumference ratio as a prognostic evaluation index, along with the BCS, for assessing patellar dislocation and other hyperlipidemia-related diseases and states such as hypertriglyceridemia and obesity-related orthopedic disease in small dogs. Eleven client-owned dogs were selected randomly among patients that visited Bundang New York Animal Hospital, South Korea from June 2021 to August 2024. According to the results, triglycerides (TG) showed a negative correlation with thorax/thigh value ($R = -0.585$, p-value = 0.059) and a strong positive correlation with thigh circumference ($R = 0.749$, p-value = 0.008). Total cholesterol (TC) showed a strong positive correlation with thigh circumference ($R = 0.776$, p-value = 0.005), whereas the thorax/thigh value showed a negative correlation with the medial patella luxation (MPL) grade with low significance ($R = -0.343$, p-value = 0.302). These data indicate that thigh circumference can be an excellent negative indicator for hyperlipidemia and thorax/thigh value shows no correlation with medial patella luxation, which has many factors such as varus and trochlear groove. Despite the limitations of this study due to the small sample size, this pilot study is significant as it is the first trial to introduce a new indicator for monitoring hyperlipidemia at home by using a simple tape measure. Also, the author reviews molecular pathways including the ApoA-1, ApoE, and LPL genes, which are related to hyperlipidemia, to explain the results.

Keywords: thorax/thigh circumference ratio; body condition score; hyperlipidemia; obesity; patellar luxation; ApoA-1; high-density lipoprotein

1. Introduction

The body condition score was first introduced in dairy cows to evaluate their productivity [1]. It is traditionally used in the form of a five-grade or nine-grade scale, and a higher grade reflects more body fat. A score of three points on the five-grade system or five points on the nine-grade system is suitable for the healthy maintenance of cows. The BCS is also applied to other animals, including small animals such as feline and canine species. The criteria of the BCS are visibility of the rib, lumber vertebrae, and pelvic bone and discernible waist in dogs, and an ideal state of body is four to five points on the nine-grade system [2,3]. As this is a somewhat subjective element, evaluation of the BCS can vary slightly depending on the person measuring it [4], and the scores are not subdivided in detail, so it is quite difficult to monitor subtle differences [5]. Failure in monitoring subtle differences using the BCS can lead to obesity [6], hyperlipidemia [7,8], and other endocrine diseases such as Cushing's disease. Hyperlipidemia commonly results in diabetes, hypothyroidism, and Cushing's syndrome in dogs [9] as well as humans [10,11]. To date, the correlation between hyperlipidemia and orthopedic diseases has not been clearly documented in dogs, and there are only a few studies on the effects of hyperlipidemia on knee tendons in rats [12,13]. However, the anatomy of dogs and rats is very different, so the results may not be applicable to dogs [14,15]. In contrast, there are many studies on the correlation between obesity, BCS, and orthopedic diseases, and numerous studies have attempted to develop monitoring indicators and prognostic evaluation biomarkers to assist in assessing the BCS [16–19]. In humans, abdominal fat is closely related to hypertension and the waist/hip ratio, which is positively correlated with cardiovascular disease in epidemiological research [20]. Additionally, high-density lipoprotein (HDL), which is known as 'good cholesterol' that cleans fat in vessel to lower the possibility of cardiovascular disease, is reported to be negatively correlated to truncal fat and waist/hip ratio [21]. Moreover, abdominal fat can increase insulin resistance, leading to a higher risk of diabetes mellitus [22]. Therefore, in order to manage abdominal fat effectively, various indicators such as waist/hip ratio and waist/chest ratio, along with Body Mass Index (BMI), have been developed [23]. Additionally, although BMI-related abdominal fat is commonly associated with diabetes mellitus, hypertension, and cardiovascular diseases, hind leg muscle circumference has been shown to be a better indicator than BMI in respiratory diseases such as Chronic Obstructive Pulmonary Disease (COPD) [24]. In contrast, there are few biomarkers or indicators other than the BCS for monitoring obesity and related diseases even though their importance is increasing in veterinary medicine [25–27]. Moreover, in veterinary medicine, there are breed differences even in the same species, so the same BCS does not always reflect the same body fat, especially in dogs [25]. For example, Schnauzers are genetically prone to hyperlipidemia because Very-Low-Density Lipoprotein (VLDL) and chylomicrons accumulate abnormally to a large degree in this breed compared to other breeds [28]. Thus, Schnauzers may be more susceptible than other breeds to hyperlipidemia-related diseases such as pancreatitis, and it has been demonstrated that it may be related to mutation of the SPINK1 gene [29]. Therefore, the author would like to introduce a new indicator, thorax/thigh ratio in dogs, which resembles the waist/hip ratio in humans. Comprehensive monitoring of body fat in a detailed but simple way is key to managing health successfully in veterinary medicine as well as in human medicine. The significance and originality of this study lies in its development of an indicator that can be used alongside the BCS and body weight to diagnose hyperlipidemia and related diseases.

2. Materials and Methods

Blood samples were collected from eleven client-owned dogs visiting Bundang New York Animal Hospital, a general practice located in South Korea. The dogs were selected randomly among patient aged between 1 to 17 years old from different breeds and sexes who visited hospital from June 2021 to August 2024 (Table 1). All blood samples were collected from the cephalic vein of the dogs after physical examination (measuring body weight, BCS, thorax, thigh circumference, patella luxation, gait) following 6 h of fasting [30] without anesthesia or sedation and centrifuged at 14,500 RPM for 1 min, and the sera were immediately and directly analyzed using a DRI-CHEM NX500 (Fujifilm, Tokyo, Japan), a dry chemistry analyzer. The samples were measured directly without dilution at room temperature. The circumference of the thigh and thorax [31] of the dogs was measured manually using a tape measure (Figure 1). JASP (version, 0.19.1, the JASP team, Amsterdam, The Netherlands) was used for data analysis (Pearson correlation, independent sample *t*-test) and graph creation (https://jasp-stats.org, accessed on 20 August 2024). Amino acid sequence analysis was performed using Vectorbuilder (Version 2.1.92, VectorBuilder Inc., Chicago, IL, USA, https://en.vectorbuilder.com/tool/sequence-alignment.html, accessed on 20 August 2024).

Table 1. Signalment and concurrent disease of 11 dogs used in correlation and comparison study. (y: year, m: month, BCS: body condition score 1 to 9, All blood work units: mg/dL, LP: left-sided pain, RP: right-sided pain, N: normal).

#	Breed	Age	Sex	Body Weight	BCS	Thigh Circumference (L/R), cm	Thorax Circumference, cm	Thorax/Thigh Ratio	Total Cholesterol (111~312)	Triglycerides (30~133)	Patella Luxation Grade (L/R)	Gait	Concurrent Disease
1	Pug	17 y	CM	7.95 kg	6/9	19/19	45	2.37/2.37	229	104	2/2	N	Patella luxation
2	Pomeranian	3 y 6 m	CM	3.75 kg	5.5/9	13/16	36	2.76/2.25	287	101	0/0	N	None
3	Bichon Frise	1 y 3 m	SF	2.73 kg	4/9	15/12	28	1.86/2.33	120	57	1.5/2	N	Patella luxation
4	Pomeranian	6 y 6 m	CM	3.4 kg	6/9	19/19	25	1.3/1.3	500	432	2/1	N	Gallbladder sludge
5	Mix	1 y	SF	2.95 kg	4/9	14/14	26	1.86/1.86	162	45	2/1	LP	Patella luxation
6	Pomeranian	5 y	SF	3.9 kg	7.5/9	18/22	43	2.38/1.95	500	333	3.5/2.5	LP	Obesity
7	Pomeranian	3 y	SF	2.5 kg	6/9	15.5/16	22	1.46/1.375	293	111	3/2.5	N	Patella luxation
8	Pomeranian	3 y 4 m	SF	4.85 kg	5.5/9	14/14	30	2.14/2.14	338	43	2.5/2.5	N	Patella luxation
9	Bichon Frise	2 y 7 m	SF	3.6 kg	6/9	14/13	29	2.07/2.23	120	87	2/1.5	N	Patella luxation
10	Maltese	6 y 6 m	CM	5.35 kg	6/9	16/17	43	2.68/2.52	305	95	2/1	N	Patella luxation
11	Maltese	5 y 2 m	CM	5 kg	5.5/9	15.5/14.5	32.5	2.09/2.24	317	86	2.5/2.5	RP	Patella luxation

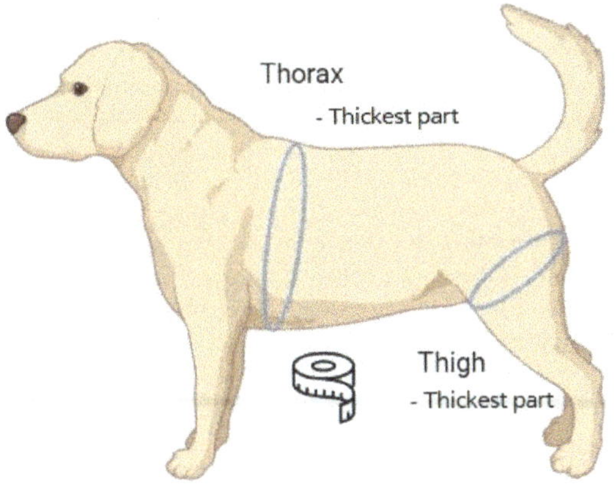

Figure 1. How to measure the thorax/thigh ratio.

3. Results

As expected, the BCS and thorax circumference showed a positive correlation (R = 0.487, p-value = 0.128), and the BCS and TC showed a significant strong positive correlation (R = 0.683, p-value = 0.021). The BCS and TG also showed a significant strong positive correlation, as expected (R = 0.587, p-value = 0.057). The BCS has been used as an indicator of obesity, and these data confirm that the BCS is indeed a valid standard.

Unfortunately, the thorax/thigh value and medial patella luxation showed a negative correlation with low significance (R = −0.343, p-value = 0.302) since patella luxation is multifactorial. The thorax/thigh value and TC showed a negative correlation (R = −0.461, p-value = 0.154). The thorax/thigh value and TG showed a negative correlation with low significance (R = −0.316, p-value = 0.344). The TG and thorax/thigh value showed a significant negative correlation (R = −0.585, p-value = 0.059). These data indicate that the thorax/thigh value has limited application at present but it shows potential, so further investigation involving a larger sample is needed.

Interestingly, total cholesterol and thigh (left side) circumference showed a positive correlation (R = 0.561, p-value = 0.073). TC and thigh (right side) circumference showed a significant strong positive correlation (R = 0.776, p-value = 0.005). Triglyceride and thigh (right side) circumference also showed a significant strong positive correlation (R = 0.749, p-value = 0.008). Triglyceride and thigh (left side) circumference also showed a significant strong positive correlation (R = 0.711, p-value = 0.014) which is consistent in both sides rather than total cholesterol (For detailed information, see the Supplementary Materials). As there are only a few studies that focused on the correlation between hyperlipidemia and thigh circumference in veterinary and human medicine, further investigation using a large sample would be beneficial to assess the utilization of thigh circumference as a prognostic indicator for hyperlipidemia.

4. Discussion

It has been revealed that although the BCS involves subjective elements, it is a reliable standard for diagnosing obesity, which is positively correlated with hyperlipidemia in dogs. To more easily monitor obesity and related diseases, the thorax/thigh circumference ratio was first introduced in this study. Although the thorax/thigh value is insignificantly correlated with medial patella luxation, as it is complex disease [32], the thorax/thigh value has a strong negative correlation with triglycerides. It can be explained that small thigh circumference may lead to an increasing level of triglycerides in humans [33], so it can be applicable to dogs, but further investigation of mechanism is needed. In fact, patellar dislocation commonly results from hindleg angular limb deformity and trochlear groove depth [34]. Therefore, the thorax/thigh ratio can be a useful indicator for hypertriglyceridemia even though there are clear limitations when applied to orthopedic diseases such as patellar dislocation and hindleg limping. Additionally, as there are many studies on the correlation between hyperlipidemia and thigh circumference in humans, validating this correlation in the veterinary field, especially in small animals, can help monitor animal health at home more effectively.

However, hyperlipidemia may be related to underlying diseases such as gallbladder sludge (Table 1, case 4, Pomeranian) or genetic predisposition (for example, idiopathic familial hyperlipidemia in Schnauzers) rather than abdominal fat or body weight in dogs, so it is important to check the underlying disease and breed difference beforehand. In fact, in cases 4 and 6, hyperlipidemia was observed in dogs with larger thighs than other dogs of similar weight (Table 1). Hindleg lipid deposition changes with aging and may be related to metabolic changes [35], implying that a large portion of the hindleg of cases 4 and 6 may consist of lipids. In addition to hindleg fat, abdominal fat is closely related to diabetes, hyperlipidemia, high blood pressure, and cardiovascular disease in humans, and there have been recent attempts to monitor abdominal fat by quantifying it with Computed Tomography (CT). The importance of abdominal fat in diseases has recently emerged in the veterinary field (Figure 2), and there are some studies that evaluated abdominal fat using CT in dogs recently [36–39].

The results of this study are significant because hyperlipidemia and obesity can be easily monitored by simply measuring body and leg circumferences, along with the BCS and body weight, at a local hospital or at home. Although this study has limitations due to the small number of dogs, the thorax/thigh ratio and thigh circumference, along with the BCS, can be secondary indicators for hyperlipidemia, which can be used to improve obesity monitoring systems to provide proper diet and supplements to prevent or treat related diseases.

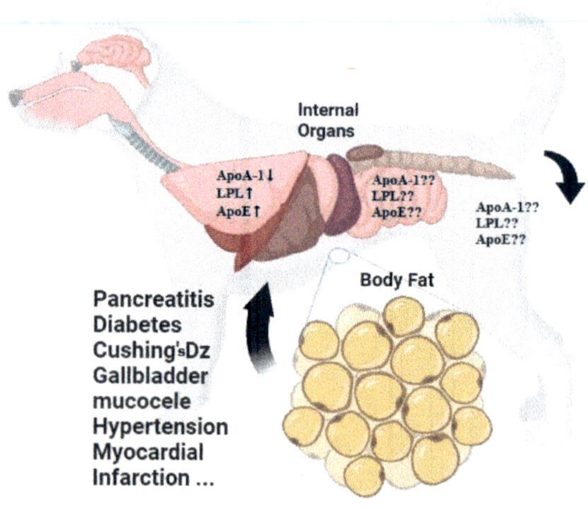

Figure 2. Schematic illustration of gene expression.

5. Reviewing Molecular Mechanism Focus on Related Genes

From the aspect of the molecular level, ApoA-1, ApoE, and lipoprotein lipase (LPL) genes are related to hyperlipidemia [40]. In particular, ApoA-1, which is known to encode a major component of HDL, may be an explanation for thigh circumference and total cholesterol correlation, because waist/hip ratio is strongly negatively associated with HDL and positive correlated with total cholesterol [41] since thigh circumference resembles the waist/hip ratio in human according to this preliminary data. There are physiological differences between humans and dogs, and thigh fat storage is strongly correlated with body weight and abdominal fat in dogs [42], which could be an interesting explanation for this pilot study's results. Also, there is only 60.9% of identity and 81.7% of similarity between the human and dog ApoE amino acid sequences (Figure 3B), which is relatively lower than human and dog ApoA1 (Figure 3A) and human and dog LPL (Figure 3C). Additionally, the ApoE gene which is known to positively correlated to cardiovascular disease and hyperlipidemia is cell-specifically expressed and mainly expressed in the liver and other organs including the small intestine and brain in humans and mice [43] (Figure 2). There is no documented estimation of ApoE mRNA concentrations in hindleg or thigh tissue in dogs; a follow up study will elucidate the fat distribution and, putting these together, it will provide clues to a deeper understanding of metabolic diseases from a comparative molecular point of view in humans and dogs. For example, we can compare human and dog ApoE mRNA concentrations in thigh fat, and investigate their comparative correlation with diabetes mellitus.

Figure 3. (**A**): Amino acid alignment of Apoa1 (Sequence 1 length:267 Sequence 2 length:266 Alignment length: 267 Identity: 227/267, 85.02%, Similarity: 246/267, 92.13%, Gaps: 1/267, 0.37%), (**B**): ApoE (Sequence 1 length:317 Sequence 2 length:323 Alignment length: 328 Identity: 226/328, 68.90%, Similarity: 268/328, 81.71%, Gaps: 16/328, 4.88%), (**C**): LPL (Sequence 1 length:475 Sequence 2 length:471 Alignment length: 478, Identity: 437/478, 91.42%, Similarity: 455/478, 95.19%, Gaps: 10/478,2.09%, sequence 1 human, sequence 2 dog).

Moreover, LPL activity is known to be positively correlated with hindleg exercise in dogs [44]. LPL is highly expressed in coronary artery disease patients [40] and a variant of LPL is closely related to hyperlipidemia [45]. Also, ApoE limits LPL-mediated hydrolysis of triglycerides [46] (Figure 4). Therefore, monitoring hindleg circumference at home after exercise has a plausible capacity to reflect hyperlipidemia blood profile.

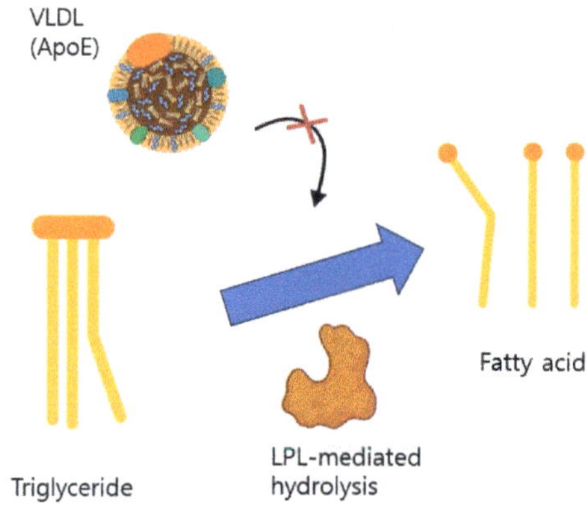

Figure 4. Illustration of how ApoE and LPL affect lipid metabolism.

Genetic mutation and polymorphism on ApoA-1, ApoE, and LPL is also closely related to hyperlipidemia [47] according to previous research by Baroni, 2003. Although this research focused on cardiovascular disease, the authors described that mutation of those genes may disturb the metabolic pathway of lipids and lead to hyperlipoproteinemia. Even after that, there are many documents which have analyzed correlation between hyperlipidemia and mutation of apolipoprotein until now, 2024 [48–50]. To elucidate the molecular pathway of hyperlipidemia in dogs, which includes many factors such as various associated genes and polymorphism and tissue-specific expression, the author recommends the single-cell RNA profiling technique which is generally used in human metabolic disease these days [51], since there is no article published where the technique is applied to hyperlipidemia dogs. This may reveal the genetic pathway of apolipoproteins organ-specifically as in the schematic illustrations shown in Figures 2 and 4.

6. Conclusions

The author briefly reviewed the molecular pathway of hyperlipidemia in dogs through pilot study data. The data showed that triglycerides showed a strong positive correlation with thigh circumference in both sides consistently (left, R = 0.711, *p*-value = 0.014, right, R = 0.749, *p*-value = 0.008) and total cholesterol in only one side (right, R = 0.776, *p*-value = 0.005), even though thorax/thigh circumference ratio failed to be a new indicator comparable to the waist/hip circumference ratio in humans. The ApoE gene may be crucial for storage of fat in the thigh since it limits LPL-mediated hydrolysis of triglycerides and LPL activity is positively correlated to thigh exercise which can raise metabolic rate and help with metabolic disease. A follow up study utilizing the single-cell RNA technique would elucidate the hyperlipidemia molecular pathway of dogs by comparing human and dog ApoE mRNA concentrations at the tissue level individually.

Supplementary Materials: The following supporting information can be downloaded at: https://www.mdpi.com/article/10.3390/life14111441/s1, Supplementary Datasets S1.

Funding: This research received no specific grant from any funding agency in the public, commercial, or not-for-profit sectors.

Institutional Review Board Statement: The study was approved by the Seoul National University Institutional Review Board (protocol code SNUIRB-2024-NH-007 and date of approval 5 August 2024).

Informed Consent Statement: Informed consent was waived (or exempted) by the IRB due to the retrospective design of this study. Prior to sample collection, all pet owners signed an informed consent form.

Data Availability Statement: Data supporting this study are included within the article and/or Supporting Materials.

Conflicts of Interest: The author declares no conflict of interest.

Abbreviations

BCS	Body Condition Score
TC	Total Cholesterol
TG	Triglycerides
MPL	Medial Patellar Luxation
HDL	High-Density Lipoprotein

References

1. Wildman, E.; Jones, G.; Wagner, P.; Boman, R.; Troutt, H., Jr.; Lesch, T. A dairy cow body condition scoring system and its relationship to selected production characteristics. *J. Dairy Sci.* **1982**, *65*, 495–501. [CrossRef]
2. Verbrugghe, A. Epidemiology of small animal obesity. In *Obesity in the Dog and Cat*; CRC Press: Boca Raton, FL, USA, 2019; pp. 1–13.
3. Dorsten, C.M.; Cooper, D.M. Use of body condition scoring to manage body weight in dogs. *J. Am. Assoc. Lab. Anim. Sci.* **2004**, *43*, 34–37.
4. Gille, S.; Fischer, H.; Lindåse, S.; Palmqvist, L.; Lärka, J.; Wolf, S.; Penell, J.; Söder, J. Dog owners' perceptions of canine body composition and effect of standardized education for dog owners on body condition assessment of their own dogs. *Vet. Sci.* **2023**, *10*, 447. [CrossRef] [PubMed]
5. Yam, P.S.; Naughton, G.; Butowski, C.F.; Root, A.L. Inaccurate assessment of canine body condition score, bodyweight, and pet food labels: A potential cause of inaccurate feeding. *Vet. Sci.* **2017**, *4*, 30. [CrossRef] [PubMed]
6. German, A.; Lequarre, A.-S.; Fredholm, M.; Lohi, H.; Häggström, J.; Willesen, J.L.; Gouni, V.; Höglund, K.; Farnir, F.; Peeters, D. Fasting triglyceride concentrations are associated with markers of lipid metabolism and glucose homeostasis in healthy, non-obese dogs in lean and overweight condition. *Front. Vet. Sci.* **2024**, *11*, 1406322.
7. Radakovich, L.B.; Truelove, M.P.; Pannone, S.C.; Olver, C.S.; Santangelo, K.S. Clinically healthy overweight and obese dogs differ from lean controls in select CBC and serum biochemistry values. *Vet. Clin. Pathol.* **2017**, *46*, 221–226. [CrossRef]
8. José Lahm Cardoso, M.; Fagnani, R.; Zaghi Cavalcante, C.; de Souza Zanutto, M.; Júnior, A.Z.; Holsback da Silveira Fertonani, L.; Calesso, J.R.; Melussi, M.; Pinheiro Costa, H.; Yudi Hashizume, E. Blood pressure, serum glucose, cholesterol, and triglycerides in dogs with different body scores. *Vet. Med. Int.* **2016**, *2016*, 8675283. [CrossRef]
9. Xenoulis, P.G.; Steiner, J.M. Lipid metabolism and hyperlipidemia in dogs. *Vet. J.* **2010**, *183*, 12–21. [CrossRef]
10. Liberopoulos, E.N.; Elisaf, M.S. Dyslipidemia in patients with thyroid disorders. *Hormones* **2002**, *1*, 218–223. [CrossRef]
11. Yadav, D.; Pitchumoni, C. Issues in hyperlipidemic pancreatitis. *J. Clin. Gastroenterol.* **2003**, *36*, 54–62. [CrossRef]
12. Grewal, N. The Link Between Hypercholesterolemia and Tendon Pathology. Ph.D. Thesis, University of British Columbia, Vancouver, BC, Canada, 2013.
13. Oniki, Y.; Kato, T.; Irie, H.; Mizuta, H.; Takagi, K. Diabetes with hyperlipidemia: A risk factor for developing joint contractures secondary to immobility in rat knee joints? *J. Orthop. Sci.* **2005**, *10*, 221–226. [CrossRef] [PubMed]
14. Hildebrand, C.; Öqvist, G.; Brax, L.; Tuisku, F. Anatomy of the rat knee joint and fibre composition of a major articular nerve. *Anat. Rec.* **1991**, *229*, 545–555. [CrossRef] [PubMed]
15. Olah, T.; Cai, X.; Michaelis, J.C.; Madry, H. Comparative anatomy and morphology of the knee in translational models for articular cartilage disorders. Part I: Large animals. *Ann. Anat. Anat. Anz.* **2021**, *235*, 151680. [CrossRef] [PubMed]
16. Kleine, S.A.; Sanderson, S.L.; George, C.; Roth, I.; Gogal, R.M.; Thaliath, M.A.; Budsberg, S.C. Correlation of serum and synovial leptin concentrations with body condition scores in healthy and osteoarthritic dogs. *Vet. Surg.* **2019**, *48*, 780–785. [CrossRef] [PubMed]
17. Leclerc, L.; Thorin, C.; Flanagan, J.; Biourge, V.; Serisier, S.; Nguyen, P. Higher neonatal growth rate and body condition score at 7 months are predictive factors of obesity in adult female Beagle dogs. *BMC Vet. Res.* **2017**, *13*, 1–13. [CrossRef]
18. Kleine, S.; Sanderson, S.; George, C.; Roth, I.; Gogal, R.; Thaliath, M.; Budsberg, S. Correlation of synovial fluid leptin with body habitus in the healthy and osteoarthritic canine knee. *Osteoarthr. Cartil.* **2019**, *27*, S473–S474. [CrossRef]
19. Webb, T.L.; du Plessis, H.; Christian, H.; Raffan, E.; Rohlf, V.; White, G.A. Understanding obesity among companion dogs: New measures of owner's beliefs and behaviour and associations with body condition scores. *Prev. Vet. Med.* **2020**, *180*, 105029. [CrossRef]
20. Björntorp, P. Abdominal fat distribution and disease: An overview of epidemiological data. *Ann. Med.* **1992**, *24*, 15–18. [CrossRef]
21. Ostlund, R.E., Jr.; Staten, M.; Kohrt, W.M.; Schultz, J.; Malley, M. The ratio of waist-to-hip circumference, plasma insulin level, and glucose intolerance as independent predictors of the HDL2 cholesterol level in older adults. *N. Engl. J. Med.* **1990**, *322*, 229–234. [CrossRef]

22. Després, J.-P.; Nadeau, A.; Tremblay, A.; Ferland, M.; Moorjani, S.; Lupien, P.J.; Thériault, G.; Pinault, S.; Bouchard, C. Role of deep abdominal fat in the association between regional adipose tissue distribution and glucose tolerance in obese women. *Diabetes* **1989**, *38*, 304–309. [CrossRef]
23. Fink, B.; Neave, N.; Manning, J. Second to fourth digit ratio, body mass index, waist-to-hip ratio, and waist-to-chest ratio: Their relationships in heterosexual men and women. *Ann. Hum. Biol.* **2003**, *30*, 728–738. [CrossRef] [PubMed]
24. Marquis, K.; Debigaré, R.; Lacasse, Y.; LeBlanc, P.; Jobin, J.; Carrier, G.; Maltais, F. Midthigh muscle cross-sectional area is a better predictor of mortality than body mass index in patients with chronic obstructive pulmonary disease. *Am. J. Respir. Crit. Care Med.* **2002**, *166*, 809–813. [CrossRef] [PubMed]
25. Palerme, J.S.; Silverstone, A.; Riedesel, E.; Simone, K.; Pomrantz, J. A pilot study on the effect of fat loading on the gastrointestinal tract of healthy dogs. *J. Small Anim. Pract.* **2020**, *61*, 732–737. [CrossRef] [PubMed]
26. Thengchaisri, N.; Theerapun, W.; Kaewmokul, S.; Sastravaha, A. Abdominal obesity is associated with heart disease in dogs. *BMC Vet. Res.* **2014**, *10*, 1–7. [CrossRef] [PubMed]
27. Witzel, A.L.; Kirk, C.A.; Henry, G.A.; Toll, P.W.; Brejda, J.J.; Paetau-Robinson, I. Use of a morphometric method and body fat index system for estimation of body composition in overweight and obese cats. *J. Am. Vet. Med. Assoc.* **2014**, *244*, 1285–1290. [CrossRef]
28. Xenoulis, P.G.; Suchodolski, J.S.; Levinski, M.D.; Steiner, J.M. Investigation of hypertriglyceridemia in healthy Miniature Schnauzers. *J. Vet. Intern. Med.* **2007**, *21*, 1224–1230. [CrossRef]
29. Bishop, M.A.; Xenoulis, P.G.; Levinski, M.D.; Suchodolski, J.S.; Steiner, J.M. Identification of variants of the SPINK1 gene and their association with pancreatitis in Miniature Schnauzers. *Am. J. Vet. Res.* **2010**, *71*, 527–533. [CrossRef]
30. Matsuzawa, T.; Sakazume, M. Effects of fasting on haematology and clinical chemistry values in the rat and dog. *Comp. Haematol. Int.* **1994**, *4*, 152–156. [CrossRef]
31. Chun, J.L.; Bang, H.T.; Ji, S.Y.; Jeong, J.Y.; Kim, M.; Kim, B.; Lee, S.D.; Lee, Y.K.; Reddy, K.E.; Kim, K.H. A simple method to evaluate body condition score to maintain the optimal body weight in dogs. *J. Anim. Sci. Technol.* **2019**, *61*, 366–370. [CrossRef]
32. Cashmore, R.; Havlicek, M.; Perkins, N.; James, D.; Fearnside, S.; Marchevsky, A.; Black, A. Major complications and risk factors associated with surgical correction of congenital medial patellar luxation in 124 dogs. *Vet. Comp. Orthop. Traumatol.* **2014**, *27*, 263–270.
33. Jung, K.J.; Lee, G.J.; Jee, S.H. The association between thigh circumference and lipids profile in Korean population: The Korea Medical Institute Study. *J. Lipid Atheroscler.* **2012**, *1*, 79–86. [CrossRef]
34. Malghem, J.; Maldague, B. Depth insufficiency of the proximal trochlear groove on lateral radiographs of the knee: Relation to patellar dislocation. *Radiology* **1989**, *170*, 507–510. [CrossRef] [PubMed]
35. Neumann-Haefelin, C.; Kuhlmann, J.; Belz, U.; Kalisch, J.; Quint, M.; Gerl, M.; Juretschke, H.P.; Herling, A.W. Determinants of intramyocellular lipid concentrations in rat hindleg muscle. *Magn. Reson. Med.* **2003**, *50*, 242–248. [CrossRef] [PubMed]
36. Kim, D.; Noh, D.; Oh, T.; Lee, K. Body fat assessment by computed tomography and radiography in normal Beagle dogs. *J. Vet. Med. Sci.* **2018**, *80*, 1380–1384. [CrossRef] [PubMed]
37. Kobayashi, T.; Koie, H.; Kusumi, A.; Kitagawa, M.; Kanayama, K.; Otsuji, K. Comparative investigation of body composition in male dogs using CT and body fat analysis software. *J. Vet. Med. Sci.* **2014**, *76*, 439–446. [CrossRef] [PubMed]
38. Turner, R.; Tyrrell, D.; Hepworth, G.; Dunshea, F.; Mansfield, C. Compartmental fat distribution in the abdomen of dogs relative to overall body fat composition. *BMC Vet. Res.* **2020**, *16*, 104. [CrossRef]
39. Nagao, I.; Ohno, K.; Nagahara, T.; Yokoyama, N.; Nakagawa, T.; Fujiwara, R.; Yamamoto, K.; Goto-Koshino, Y.; Tomiyasu, H.; Tsujimoto, H. Evaluation of visceral fat mass in dogs by computed tomography. *J. Vet. Med. Sci.* **2019**, *81*, 1552–1557. [CrossRef]
40. Dergunov, A.D.; Nosova, E.V.; Rozhkova, A.V.; Vinogradina, M.A.; Baserova, V.B.; Popov, M.A.; Limborska, S.A.; Dergunova, L.V. Differential Expression of Subsets of Genes Related to HDL Metabolism and Atherogenesis in the Peripheral Blood in Coronary Artery Disease. *Curr. Issues Mol. Biol.* **2023**, *45*, 6823–6841. [CrossRef]
41. Gillum, R. Distribution of waist-to-hip ratio, other indices of body fat distribution and obesity and associations with HDL cholesterol in children and young adults aged 4–19 years: The Third National Health and Nutrition Examination Survey. *Int. J. Obes.* **1999**, *23*, 556–563. [CrossRef]
42. Payan-Carreira, R.; Martins, L.; Miranda, S.; Olivério, P.; Silva, S.R. In vivo assessment of subcutaneous fat in dogs by real-time ultrasonography and image analysis. *Acta Vet. Scand.* **2016**, *58*, 11–18. [CrossRef]
43. Kockx, M.; Traini, M.; Kritharides, L. Cell-specific production, secretion, and function of apolipoprotein E. *J. Mol. Med.* **2018**, *96*, 361–371. [CrossRef] [PubMed]
44. Budohoski, L. Exercise-induced changes in lipoprotein lipase activity (LPLA) in skeletal muscles of the dog. *Pflügers Arch.* **1985**, *405*, 188–192. [CrossRef] [PubMed]
45. Gagne, E.; Genest, J., Jr.; Zhang, H.; Clarke, L.; Hayden, M. Analysis of DNA changes in the LPL gene in patients with familial combined hyperlipidemia. *Arterioscler. Thromb. J. Vasc. Biol.* **1994**, *14*, 1250–1257. [CrossRef] [PubMed]
46. Whitacre, B.E.; Howles, P.; Street, S.; Morris, J.; Swertfeger, D.; Davidson, W.S. Apolipoprotein E content of VLDL limits LPL-mediated triglyceride hydrolysis. *J. Lipid Res.* **2022**, *63*, 100157. [CrossRef] [PubMed]
47. Baroni, M.G.; Berni, A.; Romeo, S.; Arca, M.; Tesorio, T.; Sorropago, G.; Di Mario, U.; Galton, D.J. Genetic study of common variants at the Apo E, Apo AI, Apo CIII, Apo B, lipoprotein lipase (LPL) and hepatic lipase (LIPC) genes and coronary artery disease (CAD): Variation in LIPC gene associates with clinical outcomes in patients with established CAD. *BMC Med. Genet.* **2003**, *4*, 1–7. [CrossRef]

48. Liu, Y.; Dai, S.; Qin, S.; Zhou, J.; Wang, Z.; Yin, G. The pathogenic mutations of APOA5 in Chinese patients with hyperlipidemic acute pancreatitis. *Lipids Health Dis.* **2024**, *23*, 44. [CrossRef]
49. Medeiros, A.M.; Alves, A.C.; Miranda, B.; Chora, J.R.; Bourbon, M.; Rato, Q.; Gomes, A.C.; Ferreira, A.C.; Gaspar, A.; Marques, A.M. Unraveling the genetic background of individuals with a clinical familial hypercholesterolemia phenotype. *J. Lipid Res.* **2024**, *65*, 100490. [CrossRef]
50. Majeed, A.; Baig, Z.A.; Rashid, A. Identification and Molecular Simulation of Genetic Variants in ABCA1 Gene Associated with Susceptibility to Dyslipidemia in Type 2 Diabetes. *Int. J. Mol. Sci.* **2024**, *25*, 6796. [CrossRef]
51. Bondareva, O.; Rodríguez-Aguilera, J.; Oliveira, F.; Liao, L.; Rose, A.; Gupta, A.; Singh, K.; Geier, F.; Schuster, J.; Boeckel, J.-N.; et al. Single-cell profiling of vascular endothelial cells reveals progressive organ-specific vulnerabilities during obesity. *Nat. Metab.* **2022**, *4*, 1591–1610. [CrossRef]

life

MDPI

Article

Computed Tomography and Fluorescence Spectroscopy Blood Plasma Analysis Study for Kynurenic Acid as a Diagnostic Approach to Chronic Coenurosis in Sheep

Loredana Elena Olar [1,2,†], Vasile Daniel Tomoiagă [1,†], Sorin Marian Mârza [1,*], Ionel Papuc [1], Ioan Florin Beteg [1], Petru Cosmin Peștean [1], Mihai Musteață [3], Caroline Maria Lăcătuș [1], Raluca Marica [1], Paula Maria Pașca [3], Robert Cristian Purdoiu [1] and Radu Lăcătuș [1]

1 Faculty of Veterinary Medicine, University of Agricultural Sciences and Veterinary Medicine, 400372 Cluj-Napoca, Romania; loredana.olar@usamvcluj.ro (L.E.O.); cosmin.pestean@usamvcluj.ro (P.C.P.); raluca.marica@usamvcluj.ro (R.M.)
2 Life Sciences Institute, 3-5 Manastur Avenue, 400372 Cluj-Napoca, Romania
3 Clinics Department, Faculty of Veterinary Medicine, University of Agricultural Sciences and Veterinary Medicine, 700490 Iasi, Romania
* Correspondence: sorin.marza@usamvcluj.ro
† These authors contributed equally to this work.

Abstract: Coenurosis is a parasitic disease caused by the larval stage of *Taenia multiceps*, *Coenurus cerebralis*, typically found in the central nervous system of different livestock such as sheep and goats. The blood plasma from fifteen clinically healthy sheep and six sheep with neurological symptoms was studied by fluorescence spectroscopy in order to establish the contribution of kynurenic acid (KYNA), the neuroprotective metabolite of the kynurenine pathway, to the total fluorescence of the plasma. CT scans were used to confirm the presence of cysts in the central nervous system of sheep with neurological symptoms. The fluorescence spectroscopy analysis and further spectra deconvolution process revealed some significantly lower KYNA contributions to the total plasma fluorescence in sheep with coenurosis compared to healthy controls. Our results indicate that KYNA emission parameters could serve as valuable diagnostic markers, particularly for detecting preclinical cases of coenurosis, thus allowing for improved farm management practices.

Keywords: coenurosis; plasma; kynurenic acid; fluorescence spectroscopy

1. Introduction

Coenurosis is one of the most widespread neuropathological diseases in sheep, cattle, and goats, caused by *Coenurus cerebralis*, the larval stage of *Taenia multiceps* that inhabits the small intestine of definitive host such as dogs, wolves, and foxes. Transmission may occur via non-dewormed dogs that spread the adult tapeworm eggs on pasture [1]. According to some recent studies, the prevalence of coenurosis worldwide varies from 1.3% to 28.2%; a higher prevalence was reported in Italy, whereas in other European countries, such as Ireland, the UK, France and Greece, the rate of infections is lower [1,2]. In Romania, the prevalence of this disease is not yet established; studies are still needed in this area. Currently in Romania, there are still cases of sheep being diagnosed with coenurosis using the anamnesis and clinical manifestations, and confirmed by the post-mortem findings. Two clinical forms of coenurosis were reported in sheep—the acute and chronic conditions. Broadly speaking, the chronic form was more often described compared to the acute form [2]. Clinical signs of acute coenurosis depend on the number of migrating oncospheres in the CNS and are related to a toxic and allergic reaction rather than to a mechanical action of the oncospheres [1]. Furthermore, the clinical signs in chronic form are the consequence of the development of *Coenurus* in the cerebrum, cerebellum, or spinal cord [3,4]. Therefore, a variety of neurological symptoms were described in sheep and goats such as ataxia, incoordination, drowsiness, paralysis, muscle weakness, head compression, rotation, blurred vision, blindness and lack of a direct light reflex in the pupil, bruxism, poor appetite, seizures, and coma [1,2]. The best diagnostic method for cerebral coenurosis involves the interpretation of clinical signs with accurate localization of the cyst using diagnostic imaging techniques [1,5]. However, imaging techniques play an important role in confirming the diagnosis of coenurosis but remain expensive; thus, the need for reliable and less expansive tools is mandatory. Currently, spectroscopy techniques are being used in many medical subspecialties as an essential diagnostic tool that can help in designing and interpreting clinical and therapeutic trials [6–9]. Fluorescence spectroscopy is a widespread phenomenon because of the autofluorescence properties of various biological fluids and tissues of the body [6]. Body fluids and tissue fluorophores are produced in excess or decreased as a result of a certain pathology [6]. One of the fluorophores whose contribution in biological fluids can suffer some changes as a result of the presence of some neurological disorders is the neuroprotective metabolite of the kynurenine pathway (KP) produced by astrocytes and neurons, kynurenic acid (KYNA) [10–12]. The degradation of tryptophan is possible via the KP, which can be regulated by stress and immunocytokines (Figure 1) [13]. The KP is responsible for the neosynthesis of KYNA and picolinic, and quinolinic acids (QUIN) [11]. The two main metabolites of KP, KYNA, and QUIN are recognized as having an antagonistic effect [14]. Therefore, while KYNA is known as a neuroprotective and anticonvulsant metabolite, QUIN is recognized for its pathological effects, in particular for its ability to model neuropathological, neurochemical, and clinical features of human neurological diseases (i.e., Huntington's disease and temporal lobe epilepsy) [14–16].

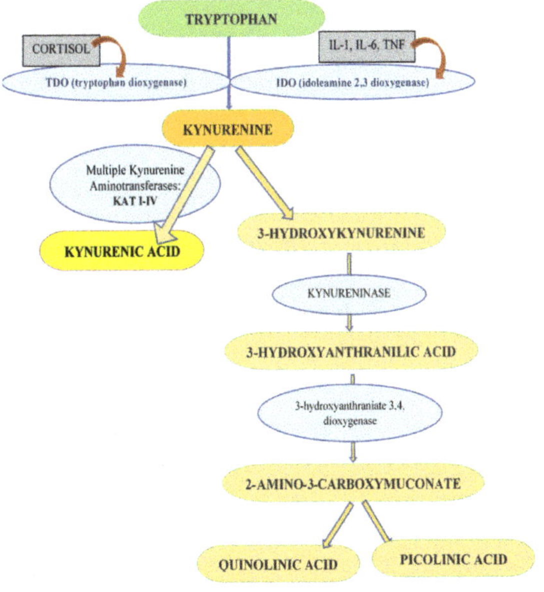

Figure 1. The catabolism of tryptophan via the kynurenine pathway. Yellow represents one of the most studied metabolites of kynurenine, kynurenic acid, which is assayed in the presented study. The enzymes responsible for the formation of kynurenic acid from kynurenine are called kynurenine aminotransferase (KAT I-IV).

The study aims to investigate the potential of KYNA as a diagnostic marker for chronic coenurosis in sheep, with emphasis on its practical application in early-stage diagnosis at the farm level. Firstly, we evaluated the clinical manifestation of the sheep with coenurosis and performed computed tomography (CT) in order to determine the location of the cysts. Secondly, in order to better understand the KYNA pathway neurobiology in health and coenurosis, we evaluated the blood plasma samples through fluorescence spectroscopy to investigate both the qualitative and quantitative contribution of KYNA to the total plasma samples in clinically healthy sheep and in sheep with chronic coenurosis confirmed by CT.

2. Materials and Methods

2.1. Animals

This study was conducted on a flock of 21 sheep belonging to a meat sheep flock reared in a semi-intensive production system. Six sheep aged between 6 months and 2 years old (two males and four females) with symptoms of head pressing, unstable gait, muscle weakness, loss of herd instinct, depression lethargy and bilateral reduction in the menace reaction were referred to the Veterinary Hospital of University of Agricultural Sciences and Veterinary Medicine of Cluj-Napoca from January 2022 to January 2024. Fifteen clinically healthy sheep (ten females and five males) from the same group were used as a control group, and their ages ranged from 6 months to 3 years. In clinical examinations, vital signs were in normal ranges, while neurological response and reflexes were decreased in the affected cases. The groups were clearly defined as "A—without neurological symptoms" and "B—with neurological symptoms" to facilitate comparison in the study. The owner stated that the flock was vaccinated and dewormed annually in accordance with the comprehensive general action plan established by the National Sanitary Veterinary and Food Safety Authority. No treatment was applied to the animals before their arrival at the hospital. Blood samples were taken from the jugular veins of the sheep and transferred into EDTA/K2-containing tubes to further perform the spectrofluorimetric analysis.

2.2. The Fluorescence of Blood Plasma Samples

The spectrofluorometric analysis of the blood plasma samples was conducted at the spectroscopy laboratory of the Life Sciences Institute "King Michael I of Romania" in Cluj-Napoca. The fluorescence spectra of the plasma samples were recorded by using a fluorescence spectrophotometer (Jasco FP-8200 ABL&E-JASCO, Budapest, Hungary). A quantity of 1 mL of blood plasma was analyzed in the 1 cm cuvette of the instrument. The fluorescence of KYNA was measured with an excitation of 343 nm and emission of 388 nm [10]. The obtained spectra were exported in OriginPro Version 8.5.1 Software (Origin, OriginLab, Northampton, MA, USA) for peak reading and further for the deconvolution process of the bands [8,17,18]. Therefore, with the peak analyzer option of the Origin Software, we fitted the peaks and identified the band associated with the presence of KYNA in the blood plasma. After each deconvolution, a report was generated with detailed information about the fitting process and each peak, including the peak at ~388 nm associated with the emission of KYNA.

2.3. Computed Tomography (CT)

CT examination of the head was performed in sternal recumbency, using a multidetector 16-slice CT scanner (Siemens Somatom Scope, Munich, Germany) on helical scan mode with the following parameters: 130 kVp, 200 mAs, 1.25 slice thickness. Pre- and post-contrast delayed-phase examination was performed using a power injector (Mallinckrodt LF Dual Head CT Injector Optistat with OptiBolus, Dublin, Ireland) to infuse a bolus of 1.5 mL/kg of iodinated non-ionic contrast medium (Iohexol, Omnipaque, 350 mg L/mL, GE Healthcare, Oslo, Norvegia). Images were acquired in a soft tissue reconstruction algorithm 45 s after the contrast medium. CT of the sheep's head was taken in axial view, and a multiplanar reconstruction was obtained in the image post-processing stage.

2.4. Statistical Analysis

A statistical analysis was conducted using IBM SPSS 29 Statistics. We conducted descriptive statistics to determine the mean and standard deviation for each parameter associated with the ~388 nm band, corresponding to the emission of KYNA.

An independent samples *t*-test was used to compare the parameters of the KYNA emission peak between control sheep and sheep with coenurosis. All results were considered statistically significant at a p value < 0.05.

3. Results

3.1. Fluorescence Spectroscopy Plasma Analysis

The fluorescence spectra of blood plasma samples collected from control sheep and, respectively, sheep with neurological symptoms associated with coenurosis are presented in Figure 2A,B. Therefore, the area of the band at 388 nm, associated with the emission of KYNA, in the spectra of the plasma samples from the sheep with clinical signs of coenurosis is modified compared with the corresponding profile of the control. Further, after the process of decomposing the peaks, we obtained the relative contribution of KYNA to the total fluorescence of plasma in clinically healthy sheep and sheep with coenurosis (Figure 2C,D). The results of the process of decomposing peaks in the spectra of the control and experimental groups of sheep concerning the band at ~388 nm are presented in Table 1.

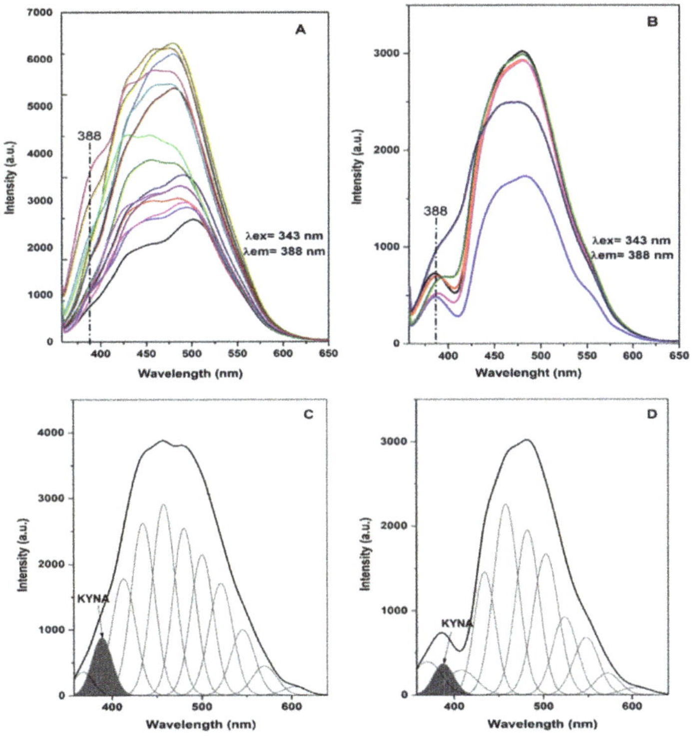

Figure 2. The fluorescence spectra of the control (**A**) and experimental groups (**B**) of sheep blood plasma samples. Deconvolution of the plasma fluorescence spectra belonging to a control sheep (**C**) and, respectively, to a sheep with neurological symptoms associated with coenurosis (**D**).

Therefore, the results show that the contribution of the KYNA to the total fluorescence of the blood plasma is significantly higher in the control group compared to the experimental group of sheep, as the descriptive statistics confirmed (Table 2).

Table 1. The results of the deconvolution process for the peak centered at ~388 nm associated with the emission of KYNA in the group of sheep without neurological symptoms (**A**) (n = 15) and with neurological symptoms (**B**) (n = 6).

AREA INT		FWHM		MAX HEIGHT		CENTER GRVTY		AREA INT P	
Group A	Group B	Group A	Group B	Group A	Group B	Group A	Group B	Group A	Group B
18,785	10,784	32	27	558	371	389	388	5.33	3.09
20,895	99,76	28.29	25	697	363	389	389	4.75	2.89
22,922	84,78	26.94	25	801	319	389	389	5.65	4.13
16,693	10,530	25.55	26	625	381	389	388	4.10	3.15
24,923	11,599	26	25	881	436	389	388	4.73	3.26
20,239	16,976	25	24	761	655	389	389	4.13	4.96
29,411		30		927		389		6.18	
22,469		26		814		389		4.98	
33,046		23		1312		389		4.65	
31,583		21		1380		388		3.72	
34,364		25		1301		389		4.37	
46,428		28		1565		389		6.23	
69,379		29		2260		388		7.91	
44,726		27		1565		389		7.21	
86,257		30		2671		388		10.15	

Area int: integrated area of the emission peak; FWHM: full width at half maximum; Max Height: maximum height of the emission peak; Center Grvty: center gravity; Area Int P: proportion of integrated area.

Table 2. Mean and standard deviation of blood plasma KYNA emission parameter for the group of sheep without neurological symptoms (**A**) and with neurological symptoms (**B**).

Parameter	Group A	Group B
AREA INT	28,225.27 ± 15,778.41	11,307.17 ± 3270.98 *
FWHM	26.85 ± 3.14	25.33 ± 1.37
MAX HEIGHT	1046.87 ± 501.31	419.17 ± 141.94 *
AREA INT P	5.12 ± 1.12	3.58 ± 1.11 *

Values are expressed as mean ± standard deviation, * $p < 0.05$. Area int: integrated area of the emission peak; FWHM: full width at half maximum; Max Height: maximum height of the emission peak; Area Int P: proportion of integrated area.

The *t*-test revealed a statistically significant difference in the integrated area of the emission peak (AREA INT) between control sheep (A) and sheep with coenurosis (B) ($t = 2.838$, $p = 0.011$). The integrated area was significantly higher in control sheep, indicating elevated KYNA levels in this group.

There was no statistically significant difference in the full width at the half maximum (FWHM) of the emission peak between the two groups ($t = 1.257$, $p = 0.224$). This suggests that the width of the KYNA emission peak remains consistent regardless of the presence of coenurosis.

The maximum height of the emission peak (MAX HEIGHT) was significantly higher in control sheep compared to those with coenurosis ($t = 3.069$, $p = 0.006$). This further supports the observation that KYNA levels are elevated in healthy sheep.

The proportion of the integrated area (AREA INT P) of the emission peak relative to the total fluorescence was significantly higher in control sheep ($t = 2.729$, $p = 0.013$). This indicates that the relative contribution of KYNA to the total blood plasma fluorescence is reduced in sheep with coenurosis.

3.2. CT

CT confirmed the presence of cysts that are mostly extra axial, supratentorial, and produce a severe mass effect on the adjacent brain parenchyma and cerebral ventricular system (Figure 3A,B). The cysts are partially well defined, and the border definition is lost when overlaying or compressing the lateral ventricles. The content is homogeneous and izoattenuated compared to the CSF (Figure 3B).

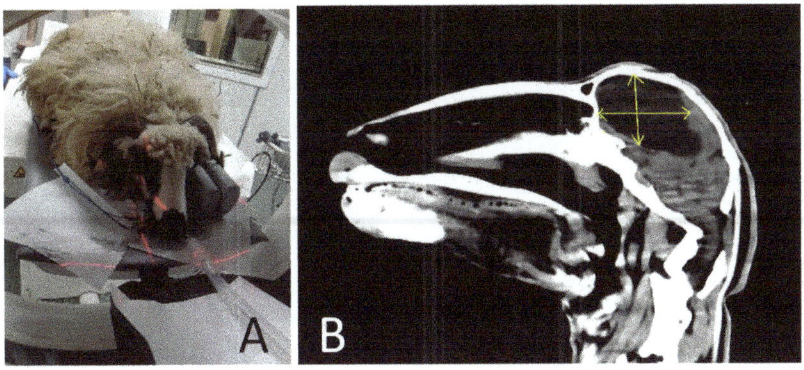

Figure 3. A CT scan of sheep (**A**). CT of a sheep brain affected by chronic coenurosis, cyst marked with yellow arrows (**B**).

4. Discussion

Currently, to the authors' knowledge, there are no studies about the spectrofluorimetric analysis of kynurenine (KYN) neuroprotective metabolites such as KYNA in blood plasma samples collected from sheep with evidence of neurological dysfunctions associated with the presence of a coenurus cyst in a ventricle or cerebral aqueduct. The diagnosis of chronic coenurosis currently is centered on relatively new imaging techniques such as CT and MRI, which were also proposed as the gold standard methods for this condition [4,19]. Our results are in agreement with previous research; CT remains a valuable non-invasive technique to diagnose encephalic conditions of sheep. Alternatively, when planning to use this modern diagnostic imaging technique in coenurosis, a clinician should also take into account the limited finances of many sheep breeders and further recommend other techniques before this one, such as fluorescence spectroscopy, which is shown to be an economical and sensitive diagnostic tool with high efficiency and sensitivity compared to many routine medical diagnostic tools for many disorders [6].

The findings from the fluorescence spectroscopy analysis indicate significant differences in several parameters of the KYNA emission peak between control sheep and those with coenurosis. Specifically, the integrated area, maximum height and proportional area of the KYNA emission peak were significantly higher in control sheep. These results suggest that KYNA levels are elevated in healthy sheep and reduced in those affected by coenurosis. The consistent peak width (FWHM) between the groups indicates that the shape of the emission peak does not vary significantly with the condition.

In recent years, several literature reports have become available on the quantitative and qualitative research of KYN metabolites in different neurological disorders in animals and humans [13,14]. The two neuroactive KP metabolites, QUIN and KYNA, were the most investigated. Therefore, studies have shown that a reduction in KYNA levels or an abnormal increase in the QUIN–KYNA ratio in the brain can lead to neuronal damage and further determine the emergence of neurological clinical manifestations [14]. However, as other authors declared, concerns exist regarding the KP metabolites, especially for KYNA and QUIN, due to the lower concentrations of those metabolites in the brain compared to those needed to model neurological diseases acutely in rodents or primates in vivo [14]. The results of the present study show that in sheep with central nervous system lesions associated with chronic coenurosis, there is a reduction in KYNA plasma levels when compared to control animals. This difference might be a consequence of the chronic neuroinflammation from coenurosis, which can activate tryptophan depletion and further enhance cerebral KP to the detriment of the serotonin pathway known as a regulator of the nervous system [13,20,21]. Hypothetically, increased KYN synthesis from tryptophan can be enhanced by mechanisms underlying neurological inflammation such as brain infiltration with circulating immune cells, activation of resident microglia and other non-neuronal cells and brain influx of blood-derived, pro-inflammatory cytokines and other immune activators [14,22,23]. Overall, therefore, our findings concerning the lower plasma contribution of KYNA to the total fluorescence of blood plasma in sheep with coenurosis compared to control sheep might suggest that once the KP is activated by the above-mentioned mechanisms of neuro-inflammation, the expression of kynurenine aminotransferase (KAT I-IV) enzymes critical for KYNA production is reduce. Decreased plasma levels of KYNA and a diminished expression of KAT have also been described in people with type 2 diabetes, comorbid to severe mental illness [24,25]. Furthermore, a disturbance within the KP with decreased levels of KYNA and increased levels of QUIN was also associated with the aggravation of a local inflammatory response and carotid artery atherosclerosis among patients with cardiovascular disease [13,26].

Therefore, the brain surplus of KYNA negatively impacts these neurotransmitter systems and further contributes to cognitive dysfunction, as signaled in a significant number of neurodegenerative diseases and psychotic disorders in humans [14,27,28]. Interestingly, in patients with schizophrenia, while brain KYNA levels are increased, the plasma levels of KYNA are decreased, highlighting the existence of an overall dissociation between the blood and central KP metabolites [24,27,28]. In our study, the peripheral findings concerning KYNA contributions in coenurosis cannot be extrapolated to the KYNA contributions in the brain or cerebrospinal fluid; however, this hypothesis needs to be tested in future studies. Hence, our findings suggest that KYNA emission parameters could be used as reliable diagnostic markers for chronic coenurosis in sheep. While the diagnostic tool detects animals with clinical symptoms, its application could extend to identifying preclinical cases at the farm level, which could lead to earlier intervention and improved management. By detecting infections before severe clinical signs develop, farmers could implement targeted deworming strategies for definitive hosts (dogs), thus reducing the spread of the disease within flocks. Future studies will explore the mechanistic pathways of KYNA in coenurosis and validate its use as a biomarker under controlled experimental conditions.

Several aspects of our study design may have influenced the results and conclusions. Notably, the study was conducted on sheep naturally infected with coenurosis, with animals at different stages of the disease and varying intensities of infection, which could have affected the test outcomes. To mitigate the impact of these variables, future studies will focus on experimentally infected sheep, allowing for a more controlled examination of KYNA's contribution to the total fluorescence of plasma across different stages of coenurosis. This approach will provide a more precise characterization of KYNA emission parameters as the disease progresses.

Another limitation of the study is the small sample size, particularly in the coenurosis group, which restricts the application of more advanced statistical techniques such as multiple analysis of variance or discriminant analysis. These methods would allow for a more comprehensive statistical assessment by considering the combined effects of multiple parameters, thereby enhancing the differentiation between infected and non-infected individuals. Future studies with larger sample sizes could greatly benefit from employing these more sophisticated statistical methods to improve the robustness of the findings.

5. Conclusions

This study provides evidence that KYNA levels in plasma may serve as diagnostic markers for chronic coenurosis in sheep. The reduction in KYNA levels in coenurosis-affected sheep could be linked to the neuroprotective role of KYNA and its involvement in neurological health. Early diagnosis facilitated by this tool could enhance management practices at the farm level by identifying infected animals before clinical symptoms become severe, allowing for timely interventions. Despite the lack of a direct treatment for sheep, the implementation of better diagnostic tools could significantly reduce the transmission of coenurosis and its economic impact on farms.

Author Contributions: Conceptualization, L.E.O. and V.D.T.; methodology, V.D.T.; software, R.C.P. and L.E.O.; validation, R.L. and I.F.B.; formal analysis, R.L.; investigation, R.M.; data curation, I.P. and I.F.B.; writing—original draft preparation, L.E.O. and M.M.; writing—review and editing, C.M.L. and P.C.P.; visualization, P.M.P.; supervision, S.M.M. All authors have read and agreed to the published version of the manuscript.

Funding: This research received no external funding.

Institutional Review Board Statement: The study was conducted in accordance with the Declaration of Helsinki and approved by the Ethics Commission of the Faculty of Veterinary Medicine in Cluj within the "University of the Agricultural Sciences and Veterinary Medicine" (No. 19637, 26 July 2024).

Informed Consent Statement: Not applicable.

Data Availability Statement: The data present in this study are available within the article.

Conflicts of Interest: The authors declare no conflicts of interest.

References

1. Varcasia, A.; Tamponi, C.; Ahmed, F.; Cappai, M.G.; Porcu, F.; Mehmood, N.; Dessì, G.; Scala, A. *Taenia multiceps* coenurosis: A review. *Parasites Vectors* **2022**, *15*, 84. [CrossRef] [PubMed]
2. Mohammadi, P.; Zakian, A.; Farjanikish, G.; Yeganeh, F.F.; Raisi, A.; Samadipoor, M. Clinical report of *Coenurosis cerebralis* outbreak in Lori sheep. *Comp. Clin. Pathol.* **2021**, *30*, 729–733. [CrossRef]
3. Scala, A.; Cancedda, G.M.; Varcasia, A.; Ligios, C.; Garippa, G.; Genchi, C. A survey of *Taenia multiceps* coenurosis in Sardinian sheep. *Vet. Parasitol.* **2007**, *143*, 294–298. [CrossRef] [PubMed]
4. Scott, P.R. Diagnosis and treatment of coenurosis in sheep. *Vet. Parasitol.* **2012**, *189*, 75–78. [CrossRef]
5. Evangelisti, M.A.; Varcasia, A.; Deiana, R.; Zobba, R.; Passino, E.S.; Scala, A.; Melosu, V.; Pipia, A.P.; Tamponi, C.; Manunta, M.L. Clinical evolution of cerebral coenurosis from invasive to chronic infec tion in sheep and a goat. *J. Infect. Dev. Ctries.* **2016**, *31*, 10. [CrossRef]
6. Shahzad, A.; Edetsberger, M.; Koehler, G. Fluorescence Spectroscopy: An Emerging Excellent Diagnostic Tool in Medical Sciences. *Appl. Spectrosc. Rev.* **2010**, *45*, 1–11. [CrossRef]
7. Musteata, M.; Nicolescu, A.; Solcan, G.; Deleanu, C. The ^{1}H NMR Profile of Healthy Dog Cerebrospinal Fluid. *PLoS ONE* **2013**, *8*, e81192. [CrossRef]
8. Ciobanu, D.M.; Olar, L.E.; Ștefan, R.; Veresiu, I.A.; Bala, C.G.; Mircea, P.A.; Roman, G. Fluorophores advanced glycation end products (AGEs)-to-NADH ratio is predictor for diabetic chronic kidney and cardiovascular disease. *J. Diabetes Its Complicat.* **2015**, *29*, 893–897. [CrossRef]
9. Olar, L.E.; Ciobanu, D.M.; Matei, F.; Papuc, I. The assessment of fluorophores advanced glycation end products-to-kynurenine ratio in healthy and diabetic rats and humans. *Stud. UBB Chem.* **2018**, *LXIII*, 37–53. [CrossRef]
10. Lim, C.; Bilgin, A.; Lovejoy, D.; Tan, V.; Bustamante, S.; Taylor, B.T.; Bessede, A.; Brew, B.J.; Guillemin, G.J. Kynurenine pathway metabolomics predicts and provides mechanistic insight into multiple sclerosis progression. *Sci. Rep.* **2017**, *7*, 41473. [CrossRef]
11. Lesniak, W.G.; Jyoti, A.; Mishra, M.M.; Louissaint, N.; Romero, R.; Chugani, D.C.; Kannan, S.; Kannan, R.M. Concurrent quantification of tryptophan and its major metabolites. *Anal. Biochem.* **2013**, *443*, 222–231. [CrossRef]
12. Vamos, E.; Pardutz, A.; Klivenyi, P.; Toldi, J.; Vecsei, L. The role of kynurenines in disorders of the central nervous system: Possibilities for neuroprotection. *J. Neurol. Sci.* **2009**, *283*, 21–27. [CrossRef] [PubMed]
13. Ala, M.; Eftekhar, S.P. The Footprint of Kynurenine Pathway in Cardiovascular Diseases. *Int. J. Tryptophan Res.* **2022**, *28*, 11786469221096643. [CrossRef]
14. Schwarcz, R.; Bruno, J.P.; Muchowski, P.J.; Wu, H.Q. Kynurenines in the mammalian brain: When physiology meets pathology. *Nat. Rev. Neurosci.* **2012**, *13*, 465–477. [CrossRef] [PubMed]
15. Schwarcz, R.; Whetsell, W.O., Jr.; Mangano, R.M. Quinolinic acid: An endogenous metabolite that produces axon-sparing lesions in rat brain. *Science* **1983**, *219*, 316–318. [CrossRef]
16. Foster, A.C.; Vezzani, A.; French, E.D.; Schwarcz, R. Kynurenic acid blocks neurotoxicity and seizures induced in rats by the related brain metabolite quinolinic acid. *Neurosci. Lett.* **1984**, *48*, 273–278. [CrossRef]
17. Muntean, C.M.; Stefan, R.; Tabaran, A.; Tripon, C.; Bende, A.; Falamas, A.; Colobatiu, L.M.; Olar, L.E. The influence of UV femtosecond laser pulses on bacterial DNA structure, as proved by Fourier transform infrared (FT-IR) spectroscopy. *ChemistrySelect* **2021**, *6*, 6957–6972. [CrossRef]

18. Muntean, C.M.; Ştefan, R.; Tăbăran, A.; Bende, A.; Fălămaş, A.; Olar, L.E. Characterization of the Structural Changes of the Genomic DNA of *Staphylococcus aureus* Due to Femtosecond Laser Irradiation by Fourier Transform Infrared (FT-IR) Spectroscopy. *Anal. Lett.* **2024**, *57*, 711–726. [CrossRef]
19. Tena, L.; De Miguel, R.; Castells, E.; Escudero, A.; Lacasta, D. Chronic coenurosis in sheep: Spontaneous remission of clinical signs and role of CT and MRI in the diagnosis and follow-up. *Vet. Rec. Case Rep.* **2020**, *8*, e001092. [CrossRef]
20. Dantzer, R. *Role of the Kynurenine Metabolism Pathway in Inflammation-Induced Depression: Preclinical Approaches;* Springer: Berlin/Heidelberg, Germany, 2017; pp. 117–138. [CrossRef]
21. Miura, H.; Ozaki, N.; Sawada, M.; Isobe, K.; Ohta, T.; Nagatsu, T. A link between stress and depression: Shifts in the balance between the kynurenine and serotonin pathways of tryptophan metabolism and the etiology and pathophysiology of depression. *Stress* **2008**, *11*, 198–209. [CrossRef]
22. Saito, K.; Markey, S.P.; Heyes, M.P. Effects of immune activation on quinolinic acid and neuroactive kynurenines in the mouse. *Neuroscience* **1992**, *51*, 25–39. [CrossRef] [PubMed]
23. Smythe, G.A.; Poljak, A.; Bustamante, S.; Braga, O.; Maxwell, A.; Grant, R.; Sachdev, P. ECNI GC-MS analysis of picolinic and quinolinic acids and their amides in human plasma, CSF, and brain tissue. *Adv. Exp. Med. Biol.* **2003**, *527*, 705–712. [CrossRef] [PubMed]
24. Campanale, A.; Inserra, A.; Comai, S. Therapeutic modulation of the kynurenine pathway in severe mental illness and comorbidities: A potential role for serotonergic psychedelics. *Prog. Progress. Neuro-Psychopharmacol. Biol. Psychiatry* **2024**, *134*, 111058. [CrossRef] [PubMed]
25. Mudry, J.M.; Alm, P.S.; Erhardt, S.; Goiny, M.; Fritz, T.; Caidahl, K.; Zierath, J.R.; Krook, A.; Wallberg-Henriksson, H. Direct effects of exercise on kynurenine metabolism in people with normal glucose tolerance or type 2 diabetes. *Diabetes Metab. Res. Rev.* **2016**, *32*, 754–761. [CrossRef]
26. Baumgartner, R.; Berg, M.; Matic, L.; Polyzos, K.P.; Forteza, M.J.; Hjorth, S.A.; Schwartz, T.W.; Paulsson-Berne, G.; Hansson, G.K.; Hedin, U.; et al. Evidence that a deviation in the kynurenine pathway aggravates atherosclerotic disease in humans. *J. Intern. Med.* **2021**, *289*, 53–68. [CrossRef]
27. Skorobogatov, K.; Autier, V.; Foiselle, M.; Richard, J.R.; Boukouaci, W.; Wu, C.L.; Raynal, S.; Carbonne, C.; Laukens, K.; Meysman, P.; et al. Kynurenine pathway abnormalities are state-specific but not diagnosis-specific in schizophrenia and bipolar disorder. *Brain Behav. Immun.-Health* **2023**, *27*, 100584. [CrossRef]
28. Han, Q.; Cai, T.; Tagle, D.A.; Li, J. Structure, expression, and function of kynurenine aminotransferases in human and rodent brains. *Cell. Mol. Life Sci.* **2010**, *67*, 353–368. [CrossRef]

life

MDPI

Article

Effect of Hot Water Bottle and Cloth Blanket on Rectal Temperature During Magnetic Resonance Imaging of the Head in Cats Under General Anesthesia

Ruxandra Pavel *, Cristina Fernoagă, Alexandru Gabriel Neagu * and Ruxandra Costea

Faculty of Veterinary Medicine, University of Agronomic Sciences and Veterinary Medicine of Bucharest, 011464 Bucharest, Romania; cristina.fernoaga@fmvb.usamv.ro (C.F.); ruxandra.costea@fmvb.usamv.ro (R.C.)
* Correspondence: ruxandra.pavel@fmvb.usamv.ro (R.P.); alexandru.neagu@fmvb.usamv.ro (A.G.N.)

Abstract: Maintaining an animal's body temperature during magnetic resonance imaging (MRI) poses great challenges, as many temperature measuring devices and warming systems are incompatible with the MRI machine. The aim of this study was to examine body temperature changes and evaluate the impact of using a hot water bottle and a cloth blanket on rectal temperature during magnetic resonance imaging in cats. We included in this study 30 cats from different breeds that underwent magnetic resonance imaging for 60 min that were randomly divided into a passively insulated group (G1) covered with a blanket ($n = 15$) and a positively heated group (G2) using a silicone hot water bottle under the abdomen and the same cloth blanket over the cat ($n = 15$). The body temperature was measured before premedication, before induction of anesthesia, and after the MRI examination. Body temperature decreased slightly but significantly ($p < 0.05$) after premedication. At the end of the MRI, body temperature had decreased more in G1 than G2 ($p = 0.033$) to 37.0 (36.5–37.5) °C and 38 (37.9–38.2) °C, respectively. This study provides clinical evidence that cats needing magnetic resonance imaging of the head can be protected from hypothermia by using a hot water bottle placed underneath their abdomen and a cloth blanket covering their full body.

Keywords: cat; body temperature; anesthesia; diagnostic imaging; hypothermia

1. Introduction

The use of MRI to diagnose neurological disorders in cats is on the rise [1]; however, the cold examination environment and the restriction on commercial active heating devices during the procedure contribute to the risk of hypothermia, as seen in dogs and rats [2–4]. For a successful MRI in humans, the patient must remain motionless to achieve high-quality images [5], which means animals require at least deep sedation or general anesthesia [6]. General anesthesia greatly impacts thermoregulation [7], often leading to hypothermia, which is the most common complication in small animal patients. Studies show that up to 84% [8,9] of these patients experience hypothermia, with the incidence reaching 64% during thoracolumbar surgery [10]. Moreover, the side-effects of anesthesia can aggravate conditions in cats with neurological disorders [1,11]. Hypothermia can result in serious, life-threatening complications, such as a higher risk of delayed recovery, wound infections (mainly observed with more pronounced hypothermia, below 35 °C, and not documented in domestic animals), coagulation disorders, and bradyarrhythmia [12,13]. Maintaining body temperature during anesthesia in patients undergoing MRI is challenging. MRI scanners rely on magnetic fields and radio waves to generate body images, requiring all electrical devices used in the scanner to be non-magnetic and electrically non-conductive. Additionally, MRI rooms are kept at low temperatures to ensure the magnet remains cool [14]. MRI rooms are typically kept at around 20 °C to operate the equipment, which can be quite cold for anesthetized animals.

A previous study demonstrated that using an insulation device made from bubble wrap and down cloth blankets could reduce the decrease in body temperature of the patients during MRI examinations that last between 30 min and 1 h. However, even with insulation, a considerable decrease in body temperature was noted during MRI examinations that lasted 2 h [15].

In this study, we evaluated a method to prevent hypothermia in cats during MRI examinations. We compared the use of a hot water bottle placed under the abdomen along with a cloth blanket to using a cloth blanket alone. Our hypothesis was that using a silicone hot water bottle together with a cloth blanket would prevent hypothermia in cats undergoing an MRI of the head.

2. Materials and Methods

The owners provided consent for their animals to be included in this study. This study was conducted on 30 feline patients, with a mean age of 4.5 years old, belonging to different breeds. In this study, cats meeting the following criteria were included: cats that needed an MRI examination of the head; those with a rectal temperature before premedication of 38 °C–39.2 °C; those with an American Society of Anesthesiologists (ASA) physical status of 2 and 3 [16], and those that had a complete MRI examination within 1 h. No animals were excluded from this study, as all met the inclusion criteria. Participants were randomly assigned to either the control group (G1) or the heat group (G2) using a computer-generated randomization sequence to minimize selection bias (www.random.org, accessed on 1 October 2024). A comprehensive and calm preanesthetic physical examination was conducted, assessing cardiovascular (electrocardiogram and pulse oximetry) and respiratory functions (counting breaths per minute by observing chest movements). Body temperature was recorded 15 min after the cats arrived at the clinic using a digital thermometer, and hydration levels were assessed through clinical signs such as skin turgor and mucous membrane moisture. We evaluated the color and moisture of mucous membranes (e.g., gingiva) to assess oxygenation and hydration. Blood samples were taken to further assess the cat's overall health, in addition to the neurological examination. We assessed baseline physiological parameters to ensure that the biochemical and hematological values would not confound the results. Blood samples were used to check for markers of systemic inflammation or metabolic imbalances, ensuring that all included cats were stable and comparable as we took into account the patients that were within the established reference ranges provided by the biochemical (BA400, BioSystems, Barcelona, Spain) and hematological (Advia 2120i, Siemens Medical, Swords, Irland) analyzer. All patients suspected to have increased intracranial pressure due to brain tumors were under specific treatment closely managed by our neurologist, treatment that included diuretics, corticosteroids, or anticonvulsants. No animals were excluded from this study, as all met the inclusion criteria.

Preparation of the animals included 6 h of fasting with free access to water before the MRI. Medication and protocols were designed to minimize risk and provide consistent conditions for comparing temperature regulation methods.

All cats were premedicated with 3 mcg/kg of dexmedetomidine (Dexdomitor® Zoetis, Orion pharma 0.5 mg/mL, Espoo, Finland) and 0.3 mg/kg of butorphanol (Butomidor® 10 mg/mL, VetViva Richter GmbH, Wels, Austria) intramuscularly (IM). The premedication protocol provided effective sedation and analgesia, minimizing patient stress and promoting a stable recovery. After premedication, a 24 G catheter was placed in the cephalic vein of all animals. Induction of anesthesia was started with a 2 mg/kg loading dose of propofol (Fresofol® 1% MCT/LCT, Fresenius Kabi Australia, Mount Kuring-gai, Australia) intravenously (IV) given in 15 s, followed by laryngeal desensitization with 0.5 mg/kg of 2% lidocaine (Xilina 2%, 2 mL, S.C. Zentiva SA, Romania, Bucharest) withdrawn with a 1 mL syringe and applied on the larynx.

We waited an additional 15 s before attempting to intubate the cats. If the patient showed resistance, we administered another 1 mg/kg bolus of propofol.

Isoflurane (Isothesia, 1000 mg/g, Piramal Critical Care B.V., Voorschoten, The Netherlands) in oxygen (500 mL + 10 mL/kg) was used to maintain anesthesia at a vaporizer setting of 2% through an anesthesia machine (AEON 7200A, Beijing, China) with the use of a rebreathing system.

End-tidal CO_2 ($EtCO_2$) was measured with a sidestream capnograph, and all cats kept their $EtCO_2$ between 35 and 45 mmHg. Blood pressure was monitored noninvasively using an oscillometric method. A 2.5 cm cuff was positioned at the base of the tail to ensure accurate measurements. The monitoring equipment (Comen, STAR8000-V, Shenzhen, China) was placed outside the MRI machine to avoid interference with the imaging system. During the MRI procedure, the mean arterial pressure (MAP) was consistently maintained within normal limits (MAP > 60 mmHg). We applied an ocular lubricant (Claromed®, 30 g, 1% hyaluronic acid, Inmed, Bucharest, Romania) to all cats to protect the cornea during the MRI examination.

The cats were randomly allocated into two groups using a computer-generated randomization sequence to reduce selection bias. The first group (G1), designated as the control group (Figure 1), was covered with a 100% polyester cloth blanket measuring 160 cm × 120 cm. The second group (G2, heat group) had a silicone hot water bottle (35 cm × 20 cm) placed under their abdomen, with their entire body also covered by the same polyester cloth blanket. The control group (G1, n = 15) had a median age of 5 (6–2) years old, belonging to different breeds represented by British Shorthair (n = 4), Bengal (n = 1), Domestic Shorthair (n = 4), Ragdoll (n = 1), Main Coon (n = 1), Scottish Fold (n = 2), and Siamese (n = 2). The positively heated group (G2, n = 15) had a median age of 5 (7–3) years old, belonging to different breeds represented by Main Coon (n = 1), British Shorthair (n = 3), Domestic Shorthair (n = 6), Angora (n = 1), Exotic Shorthair (n = 1), Persian (n = 1), Birman (n = 1), and Siamese (n = 1).

Figure 1. Patient from G1 in the recovery time after the MRI examination.

The hot water bottle was made entirely of silicone and was covered with a soft fleece to prevent burning the patient (Figure 2a). It was placed underneath the abdomen of each cat, and a cloth blanket was placed over the body as for G1 (Figure 2b).

The temperature of the water inside the bottle at the beginning of the MRI was 60 °C and that at the end was 55 (54–56) °C. The temperature was measured using a calibrated electronic kitchen thermometer (Westmark, Sauerland Region, Elspe, Germany). The volume of the water added to the bottle was 1.5 L.

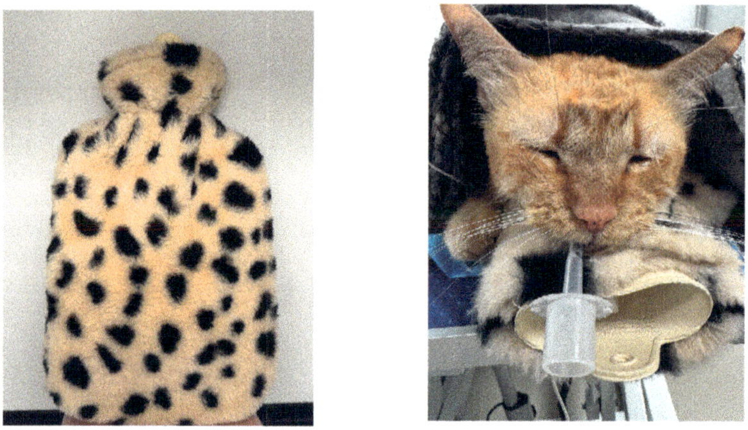

(a) (b)

Figure 2. (**a**) The hot water bottle used for this study; (**b**) position of the hot water bottle along with the cloth blanket.

The body temperature of the animals was measured rectally using a digital thermometer (Laica S.p.A., Viale del Lavoro, Italy) (which was calibrated at the beginning of each day of measurements) before premedication (15 min after they arrived in the clinic) at the time of anesthesia induction (before giving the propofol—10 min after the premedication) and at the end of the MRI examination (after 1 h). Calibration of the thermometer was represented by adding ice-cold water into a bowl, letting it rest for 2 min, and then inserting the tip of the thermometer on the surface. We took two measurements that showed 0 °C, and then we adjusted the thermometer by pressing and holding the reset button. The flexible tip of the thermometer was inserted about 2.5 cm into the patient's rectum.

Efforts were made to maintain full contact between the thermometer and the rectal mucosa to reduce any interference from fecal matter during measurement. The measurement started once the thermometer was correctly positioned in the rectum and concluded when an acoustic signal indicated completion. The temperature inside the MRI room was maintained at 20 °C. The cats were placed in the prone position with their head inside the coil. All MRI scans were acquired using high-field-strength 1.5-Tesla magnets (Magnetom Essenza, Siemens Healthineers, Shenzhen, China) with a 16-channel head and neck coil. For this study, a complete MRI series was defined as including, at minimum, sagittal, transverse, and dorsal T2-weighted images, transverse fluid-attenuated inversion recovery (FLAIR), and transverse pre-contrast and post-contrast T1-weighted images. All measurements were conducted around the same time of the day, between 4 p.m. and 7 p.m., to account for potential variations in the temperature around the day. Additionally, the temperature measurements were performed by the same individual.

All cats underwent the same duration of anesthesia, which was 1 h. Patients were closely monitored during the recovery period. Recovery time was defined as the interval from extubation to the moment the cat was able to stand unassisted. Observations were continuous, ensuring an accurate assessment of each animal's recovery progress. A statistical analysis of the data was performed on the web application DATAtab (DATAtab Team 2024. DATAtab: Online Statistics Calculator. DATAtab e.U. Graz, Austria. https://datatab.net, accessed on 12 October 2024) and Microsoft Excel (Version 16.82, 2024).

Median values, along with the range, were calculated for the temperature results. A Mann–Whitney U nonparametric statistical test was performed for the evaluation of body temperature changes in the control and heat groups, along with the differences between groups regarding duration of MRI scanning, body weight, ASA classification, anesthesia duration, and recovery time.

For more than two nominal variables, the comparison was conducted through the Kruskal–Wallis test. Statistical significance was defined as a p-value <0.05.

3. Results

There were no significant differences in body weight, time under anesthesia, or ASA classification between the two groups. The MRI diagnosis of the patients from G1 was represented by idiopathic epilepsy (5), glioma (3), meningioma (2), internal bilateral otitis (1), inflammatory process in the right bulla (1), right hemisphere tumor (1), left hemisphere posttraumatic lesion (1), and encephalitis (1). For G2, the MRI diagnosis was meningioma (4), idiopathic epilepsy (3), encephalitis (2), meningoencephalitis (2), glioma (2), interstitial edema (1), and bilateral frontal ophthalmic cyst (1).

The body temperature before premedication was 38.9 (38.7–39.0) °C in both groups (p = 0.099). The body temperatures before induction of anesthesia for G1 and G2 were 38.4 (37.8–38.7) °C and 38.4 (38.1–38.7) °C, with a p-value of 0.41. Following the MRI, the temperatures were 37.0 (36.5–37.5) °C and 38 (37.9–38.2) °C in G1 and G2, respectively, and only the value in G1 was significantly different from the value prior to the MRI (p = 0.033).

The recovery times for G1 and G2 were 17 (25–13) minutes and 11 (15–7) minutes, respectively, which were statistically significantly different (p < 0.001).

Besides the medications administered and ASA classification, the length of anesthesia is also an important factor contributing to hypothermia during anesthesia.

4. Discussion

Based on the findings of this study, the results suggested that putting a hot water bottle underneath the patient's abdomen, along with a cloth blanket that covers the entire patient, prevented any temperature decrease for the allotted period of MRI examination of the head in cats.

Hypothermia can delay recovery through multiple mechanisms. In our study, the control group exhibited a longer recovery time (measured as time to standing) than cats in the heat group. This is likely due to hypothermia's effect on reducing hepatic blood flow and enzyme activity, which slows the metabolism of anesthetic drugs. We expected that minimizing hypothermia—using a hot water bottle under the abdomen and covering the cats' bodies—would enhance recovery time and help prevent other hypothermia-related complications [17]. Patients with lower body temperatures may take longer to recover from anesthesia, as they need to eliminate a larger amount of inhalant anesthetic. However, while the solubility of inhalant agents like isoflurane and enflurane does increase with decreasing temperature, the effect is relatively small—approximately 6% per degree of temperature drop [18]. Thus, based on the temperature range observed in this study, this factor is unlikely to have a significant effect on anesthetic clearance during recovery.

The initial decline in body temperature during anesthesia is largely attributed to shifts in blood distribution triggered by anesthetic agents, as seen in humans [19], but we do not have any data regarding animals.

Small animals lose body heat faster than larger ones because they have a higher surface-area-to-mass ratio [20], which accelerates heat loss through convection, conduction, radiation, and evaporation. In this study, the drop in rectal temperature observed can be explained by several factors.

The use of a cloth blanket likely reduced heat loss through convection and conduction by insulating the animals and limiting heat transfer to the surrounding air and surfaces. Additionally, placing a hot water bottle under the abdomen warmed the air around the animal, which helped decrease heat loss through radiation by creating a localized warm zone. Decreased muscle activity due to anesthesia, along with the drugs' effects on the hypothalamus [21] and central alpha-2 adrenoceptors [22], also contributed to reduced internal heat production. Previous studies by Kuusela et al. [23], Golden et al. [24], and Ansah et al. [25] reported similar temperature drops with dexmedetomidine, propofol, and isoflurane. Although these factors played a role in the overall temperature decrease, no significant differences were found between the groups before anesthesia induction.

5. Conclusions

This study found that placing a hot water bottle under the abdomen, along with a cloth blanket covering the entire body, was more effective in maintaining rectal temperature in cats during head MRI examinations compared to using only a cloth blanket.

Author Contributions: Conceptualization, R.P., R.C., A.G.N. and C.F.; methodology R.P., R.C., A.G.N. and C.F.; software, R.P., R.C., A.G.N. and C.F.; validation, R.P., R.C., A.G.N. and C.F.; investigation, R.P., R.C., A.G.N. and C.F.; resources, R.P., R.C., A.G.N. and C.F.; data curation, R.P., R.C., A.G.N. and C.F.; writing—original draft preparation, R.P., R.C., A.G.N. and C.F.; writing—review and editing, R.P., R.C., A.G.N. and C.F. All authors have read and agreed to the published version of the manuscript.

Funding: This research received no external funding.

Institutional Review Board Statement: Ethical review and approval were waived for this study due to the written informed consent obtained from the cat owners.

Informed Consent Statement: Informed consent was obtained from the owners of all the patients involved in this study. Written informed consent was obtained from the cat owners for the publication of this paper.

Data Availability Statement: The data generated in this study are presented in this article. For any further information, the reader can contact the authors.

Acknowledgments: We acknowledge all the colleagues who helped us in completing many cases, which were the basis of our experiences necessary in writing this article.

Conflicts of Interest: The authors declare no conflicts of interest.

References

1. Daly, E.; Cardy, T.J.; McFadzean, W.; Taylor-Brown, F.E. Perianaesthetic Complications in Cats Undergoing MRI of the Brain. *J. Feline Med. Surg.* **2024**, *26*, 1098612X241275023. [CrossRef]
2. Redondo, J.E.A.; Rubio, M.; Soler, G.; Serra, I.; Soler, C.; Gómez-Villamandos, R.J. Normal Values and Incidence of Cardiorespiratory Complications in Dogs during General Anaesthesia. A Review of 1281 Cases. *J. Vet. Med. Ser. A* **2007**, *54*, 470–477. [CrossRef] [PubMed]
3. Kushikata, T.; Hirota, K.; Kotani, N.; Yoshida, H.; Kudo, M.; Matsuki, A. Isoflurane Increases Norepinephrine Release in the Rat Preoptic Area and the Posterior Hypothalamus in Vivo and in Vitro: Relevance to Thermoregulation during Anesthesia. *Neuroscience* **2005**, *131*, 79–86. [CrossRef] [PubMed]
4. Iizuka, T.; Kamata, M.; Yanagawa, M.; Nishimura, R. Incidence of Intraoperative Hypotension during Isoflurane–Fentanyl and Propofol–Fentanyl Anaesthesia in Dogs. *Vet. J.* **2013**, *198*, 289–291. [CrossRef]
5. Isaacson, D.L.; Yanosky, D.J.; Jones, R.A.; Dennehy, N.; Spandorfer, P.; Baxter, A.L. Effect of MRI Strength and Propofol Sedation on Pediatric Core Temperature Change. *J. Magn. Reson. Imaging* **2011**, *33*, 950–956. [CrossRef] [PubMed]
6. Redondo, J.; Suesta, P.; Gil, L.; Soler, G.; Serra, I.; Soler, C. Retrospective Study of the Prevalence of Postanaesthetic Hypothermia in Cats. *Vet. Rec.* **2012**, *170*, 206. [CrossRef]
7. Sessler, D.I.; McGuire, J.; Sessler, A.M. Perioperative Thermal Insulation. *Anesthesiology* **1991**, *74*, 875–879. [CrossRef]
8. Redondo, J.; Suesta, P.; Serra, I.; Soler, C.; Soler, G.; Gil, L.; Gómez-Villamandos, R. Retrospective Study of the Prevalence of Postanaesthetic Hypothermia in Dogs. *Vet. Rec.* **2012**, *171*, 374. [CrossRef]
9. Clark-Price, S.C.; Fischer, B.L.; Kirwin, K.L.; Keating, S.C.; Auckburally, A.; Flaherty, D. Multicenter Study to Investigate Factors Associated with Change in Rectal Temperature during Anesthesia in Dogs. *J. Am. Vet. Med. Assoc.* **2021**, *258*, 64–71. [CrossRef]
10. Bruniges, N.; Rioja, E. Intraoperative Anaesthetic Complications in Dogs Undergoing General Anaesthesia for Thoracolumbar Hemilaminectomy: A Retrospective Analysis. *Vet. Anaesth. Analg.* **2019**, *46*, 720–728. [CrossRef]
11. Gaynor, J.S.; Dunlop, C.I.; Wagner, A.E.; Wertz, E.M.; Golden, A.E.; Demme, W.C. Complications and Mortality Associated with Anesthesia in Dogs and Cats. *J. Am. Anim. Hosp. Assoc.* **1999**, *35*, 13–17. [CrossRef] [PubMed]
12. Armstrong, S.R.; Roberts, B.K.; Aronsohn, M. Perioperative Hypothermia. *J. Vet. Emerg. Crit. Care* **2005**, *15*, 32–37. [CrossRef]
13. Pottie, R.; Dart, C.; Perkins, N.; Hodgson, D. Effect of Hypothermia on Recovery from General Anaesthesia in the Dog. *Aust. Vet. J.* **2007**, *85*, 158–162. [CrossRef]

14. Khenissi, L.; Covey-Crump, G.; Knowles, T.G.; Murrell, J. Do Heat and Moisture Exchangers in the Anaesthesia Breathing Circuit Preserve Body Temperature in Dogs Undergoing Anaesthesia for Magnetic Resonance Imaging? *Vet. Anaesth. Analg.* **2017**, *44*, 452–460. [CrossRef]
15. Onozawa, E.; Azakami, D.; Seki, S.; Hamamoto, Y.; Ishioka, K. Effect of an Insulation Device in Preventing Hypothermia during Magnetic Resonance Imaging Examinations for Dogs and Cats under General Anesthesia. *Animals* **2021**, *11*, 2378. [CrossRef] [PubMed]
16. Hosgood, G.; Scholl, D.T. Evaluation of Age and American Society of Anesthesiologists (ASA) Physical Status as Risk Factors for Perianesthetic Morbidity and Mortality in the Cat. *J. Vet. Emerg. Crit. Care* **2002**, *12*, 9–15. [CrossRef]
17. Sakata, H.; Walsh, V.; Chambers, J.; Bridges, J.; Sano, H. Effect of Insulation with Bubble Wrap and an Absorbent Pad on Heat Loss in Anaesthetised Cats. *N. Z. Vet. J.* **2020**, *68*, 324–330. [CrossRef]
18. Lockwood, G.; Sapsed-Byrne, S.; Smith, M. Effect of Temperature on the Solubility of Desflurane, Sevoflurane, Enflurane and Halothane in Blood. *Br. J. Anaesth.* **1997**, *79*, 517–520. [CrossRef]
19. Grimm, K.A. Perioperative Thermoregulation and Heat Balance. In *Veterinary Anesthesia and Analgesia: The Sixth Edition of Lumb and Jones*; Wiley: New York, NY, USA, 2024; pp. 246–253.
20. Hill, R.C.; Scott, K.C. Energy Requirements and Body Surface Area of Cats and Dogs. *J. Am. Vet. Med. Assoc.* **2004**, *225*, 689–694. [CrossRef]
21. Virtanen, R. Pharmacological Profiles of Medetomidine and Its Antagonist, Atipamezole. *Acta Vet. Scand. Suppl.* **1989**, *85*, 29–37.
22. Sabbe, M.B.; Penning, J.P.; Ozaki, G.T.; Yaksh, T.L. Spinal and Systemic Action of the Alpha 2 Receptor Agonist Dexmedetomidine in Dogs. Antinociception and Carbon Dioxide Response. *Anesthesiology* **1994**, *80*, 1057–1072. [CrossRef] [PubMed]
23. Kuusela, E.; Vainio, O.; Kaistinen, A.; Kobylin, S.; Raekallio, M. Sedative, Analgesic, and Cardiovascular Effects of Levomedetomidine Alone and in Combination with Dexmedetomidine in Dogs. *Am. J. Vet. Res.* **2001**, *62*, 616–621. [CrossRef] [PubMed]
24. Golden, A.; Bright, J.; Daniel, G.; Fefee, D.; Schmidt, D.; Harvey, R. Cardiovascular Effects of the (~2-Adrenergic Receptor Agonist Medetomidine in Clinically Normal Cats Anesthetized with Isoflurane. *Am. J. Vet. Res.* **1998**, *59*, 509–513. [CrossRef] [PubMed]
25. Ansah, O.B.; Raekallio, M.; Vainio, O. Correlation between Serum Concentrations Following Continuous Intravenous Infusion of Dexmedetomidine or Medetomidine in Cats and Their Sedative and Analgesic Effects. *J. Vet. Pharmacol. Ther.* **2000**, *23*, 1–8. [CrossRef]

life

MDPI

Article

Ureteral Closure Using Advanced Bipolar Vessel Sealing Devices During Laparoscopic Nephrectomy in Dogs and Cats: A Pilot Series of Clinical Cases

Przemysław Prządka [1,*], Bartłomiej Liszka [1], Kamil Suliga [1], Agnieszka Antończyk [1], Zdzisław Kiełbowicz [1], Dominika Kubiak-Nowak [1], Stanisław Dzimira [2], Adam Skalski [3] and Ludwika Gąsior [1]

1 Department and Clinic of Surgery, Faculty of Veterinary Medicine, Wroclaw University of Environmental and Life Sciences, 50-366 Wroclaw, Poland; bartlomiej.liszka@upwr.edu.pl (B.L.); kamil.suliga@upwr.edu.pl (K.S.); agnieszka.antonczyk@upwr.edu.pl (A.A.); zdzislaw.kielbowicz@upwr.edu.pl (Z.K.); dominika.kubiak-nowak@upwr.edu.pl (D.K.-N.); ludwika.gasior@upwr.edu.pl (L.G.)

2 Department of Pathology, Division of Pathomorphology and Veterinary Forensics, Faculty of Veterinary Medicine, Wroclaw University of Environmental and Life Sciences, 50-366 Wroclaw, Poland; stanislaw.dzimira@upwr.edu.pl

3 University Centre of General and Oncological Surgery, Wroclaw Medical University, 50-367 Wroclaw, Poland; adam.skalski@umw.edu.pl

* Correspondence: przemyslaw.przadka@upwr.edu.pl

Abstract: Recently, laparoscopic nephrectomy has become more popular in veterinary medicine. In the majority of these procedures, vascular sealing devices (VSDs) are used. These allow for the closure of renal vessels with advanced bipolar coagulation. However, until now, closure of the ureter was performed with mechanical clips or suturing. There is a lack of information in the literature about the possibility of VSDs being used for ureter closure. This article presents the possibility of renal vessels and ureter closure in cats and dogs with vascular sealing devices. Laparoscopic nephrectomy in dogs and cats was performed entirely with VSDs. Patients with unilateral hydronephrosis qualified for the procedure. The nephrectomies were completely performed using a laparoscopic approach. Both renal vasculature and ureter were closed with VSDs. Additionally, two resected ureters from operated cats underwent histopathological evaluation. Among the operated animals, there were no postoperative complications or signs in the urinary tract. Histopathological evaluation of two cats' ureters showed lumen closure on the coagulation places. Vascular sealing devices, during laparoscopic nephrectomy, allow for closure of not only the renal vessels but also ureters.

Keywords: laparoscopy; surgery; animals; urology

1. Introduction

A variety of unilateral renal disorders in dogs and cats can be treated by performing a nephrectomy. Traditionally, nephrectomy is performed via an open surgery approach. During this procedure, after a ventral median celiotomy, the kidney is dissected from its retroperitoneal attachments. The renal artery, renal vein, and distal ureter are ligated and dissected [1]. In veterinary medicine, nephrectomy can be used as a treatment for dogs and cats with renal disorders, including ureteral atresia, renal dysplasia, idiopathic renal hematuria, hydronephrosis, primary renal neoplasia, nephrolithiasis, polycystic kidney disease, chronic or unresponsive pyelonephritis, and trauma [1–6].

The published work that can be found on transperitoneal laparoscopic nephrectomy in research models or clinical patients within veterinary literature is limited [1,7–10]. Secure maintenance of homeostasis within surgery is crucial to safety in all operations.

For mechanical coaptation of 3 to 7 mm vessels within advanced laparoscopic nephrectomy, the most commonly used method are clips [11]. However, there is the possibility for clips to become dislodged during operative manipulation. If repeated applications are needed within the same area, they may also interfere with each other or with other stapling devices [11]. Moreover, due to their iatrogenic properties as a foreign body, clips can potentially erode into the surrounding tissue, which has been reported within the literature [12–14]. An alternative solution for vessel closure is ligation with suture, which can be more demanding technically and time consuming, especially during laparoscopic procedures [15]. Another option is the use of an advanced bipolar vessel sealing device. Using a combination of electrical currents and mechanical pressure, this device fuses vessel walls and creates seals. Coagulum is formed through mechanical pressure when collagen and elastin denature occur within the targeted tissue [15,16]. Within human surgery, evidence has shown that major branches of the renal vein, including the gonadal, adrenal, and lumber veins, can be sealed using vessel sealing devices [11]. At the same time, in animals, the renal vessels can be sealed using a vessel sealing device alone however the ureter is still closed using vascular clips [17]. To our knowledge, there is no information in veterinary literature about the use of a vessel sealing device in ureteral closure during nephrectomy in dogs and cats. The aim of this study was to evaluate the effectiveness and safety of ureteral closure using only a vessel sealing device during laparoscopic nephrectomy in dogs and cats.

2. Materials and Methods

2.1. Study Design and Patient Selection

This study was retrospectively approved by the ethics committee of Wroclaw University of Environmental and Life Sciences (Faculty of Veterinary Medicine Animal Welfare Advisory Team). All animal owners signed a consent form, after an explanation about the details of anesthesia and surgery procedures and their associated risks. Complete blood cell counts and biochemistry profiling were performed before the unilateral nephrectomy procedure.

The inclusion criteria for this study were dogs and cats with unilateral obstructive hydronephrosis requiring nephrectomy (Table 1). All operated animals were referred to the Department and Clinic of Surgery, Faculty of Veterinary Medicine, Wroclaw University of Environmental and Life Sciences in the years 2017–2022. All nephrectomies were performed by the first author in accordance with the applicable recommendations [1,18]. Animals qualified for the procedure had changes limited to one kidney, with the opposite kidney functioning properly based on ultrasound abdominal examination, morphological and biochemical blood tests and urinalysis results. Clinical observations were carried out for a minimum of one year after surgery.

Table 1. Listing and additional information about operated animals.

Animal Number	Species	Breed	Sex	Age (Years)	Body Weight (kg)	Affected Kidney	Cause of Hydronephrosis
1	dog	Giant schnauzer	female	6	32	left	iatrogenic (OVH)
2	dog	Yorkshire Terrier	male	4	3.5	right	iatrogenic (ureterotomy)
3	dog	Maltese	male	8	4	left	ureteral stones
4	cat	European shorthair	male	6	3.5	left	ureteral stones
5	cat	European shorthair	female	7	4	right	ureteral stones
6	cat	European shorthair	female	5	3	left	ureteral stones
7	cat	Scottish fold cat	female	8	3.5	left	ureteral stones

2.2. Diagnostic Procedures

All animals underwent clinical and ultrasound examinations before the procedure to confirm the presence of moderate hydronephrosis with severe loss of renal parenchyma. Additionally, the ureters were examined using ultrasound from the kidney to the bladder. This allowed for measurement of the distended part cranially to the obstruction. Complete blood analysis (morphology and biochemistry), urinalysis, and urine culture were performed before surgery. Assessment of the functioning of the contralateral kidney using scintigraphy was not performed due to the lack of availability of such a test in the country (Poland).

2.3. Surgery

All procedures were performed under general anesthesia. For the nephrectomy, the animal was placed in lateral recumbency with the affected kidney facing upward. The operated animal was covered with a sterile surgical drape immediately after antiseptic preparation in the surgical field. In all cases, an open technique with a 5 mm reusable trocar (through a longitudinal incision in the umbilicus) was used for establishing pneumoperitoneum using medical CO_2. The optics were inserted after reaching an insufflation pressure in the abdominal cavity, which was 8 mmHg. Subsequently, under the control of the endoscope, two consecutive 5 mm and 10 mm diameter trocars were inserted in a triangular fashion. After gaining access to the abdominal cavity, the affected kidney was removed using a modified technique presented by Mayhew et al. [1,17]. The instruments used during the nephrectomy were Kelly dissection forceps (Karl Storz SE & Co. KG; Tuttlingen, Germany) and a vessel sealing device with an L-hook (ligasure™ retractable l-hook laparoscopic sealer/divider, Covidien, Minneapolis, MN, USA). To initiate dissection of the kidney, a vessel sealing device or monopolar electrosurgical L-hook were used to dissect the kidney from its retroperitoneal attachments. Next, the ureter was dissected out close to its insertion into the renal pelvis, which aided in the identification of the renal hilus by providing mild traction of the ureter. After renal vessels visualization and preparation, they were sealed and divided using the vessel sealing device alone. The remaining attachments of the kidney to the surrounding retroperitoneum were then sectioned using the vessel sealer. After the kidney was completely dissected, tension was placed on the proximal ureter. The ureter was cut distally beyond the level of the obstruction causing the hydronephrosis. Ureteral obstructions were identified during surgery based on macroscopic appearance—distention of the ureter cranially to the obstruction and its narrowing caudally to the obstruction. A vessel sealing device (VSD-ligasure™ retractable l-hook laparoscopic sealer/divider, Covidien, Minneapolis, MN, USA) was used to close the lumen of the ureter. In order to close the lumen of the ureter, the ureter was grasped in the VSD forceps, and the device was activated. After the completed coagulation cycle, the ureter was released without cutting the coagulated ureter tissues. Then, the same procedure was repeated 3–5 mm proximal to the previous coagulation site. After the completed coagulation cycle, the coagulated ureter tissues were cut with a knife present in the jaws of the VSD (Figure 1A–D). To facilitate resected kidney removal, its volume was reduced by puncturing with a needle introduced through the skin and collecting fluid within a syringe. Then, the resected specimen was placed into a specimen retrieval bag and removed through an enlargement of one of the instrument ports. The entire surgical site was re-examined for any ongoing hemorrhage and CO_2 from the peritoneal cavity was removed. The trocar wound was closed using single intermittent sutures (monofilament 2/0; Dafilon; B. Braun, Rubi, Spain).

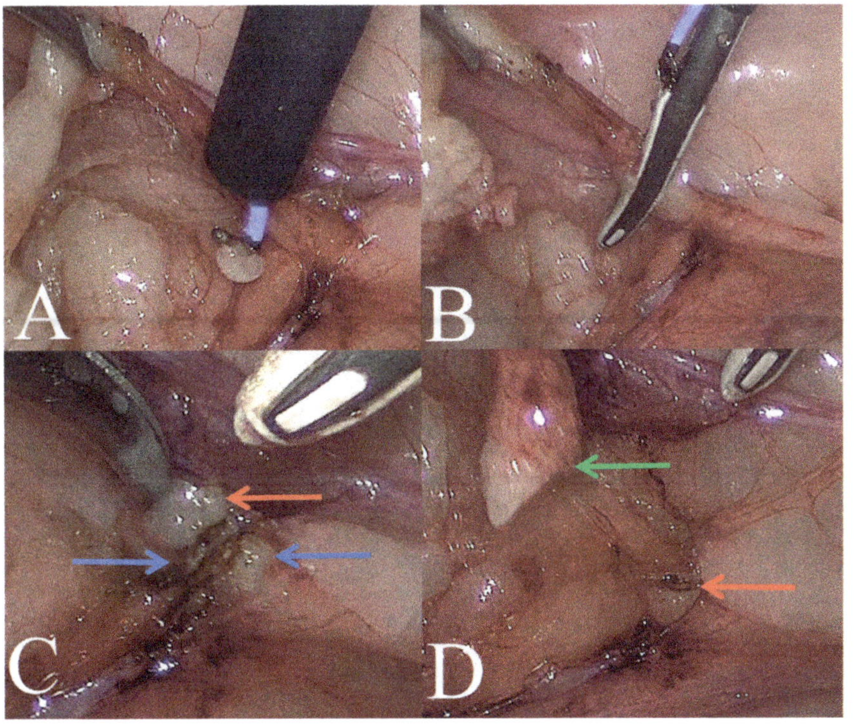

Figure 1. Intraoperative view of ureteral occlusion using a vessel sealing device (VSD) in a cat. (**A**)—dissection of the ureter; (**B**)—closing the ureter with a VSD; (**C**,**D**)—view of the ureter stump after its second closure and cutting using a VSD (green arrow—proximal part of the ureter, red arrow—ureteral incision line and the site of its closure with a VSD, blue arrow—distal part of the ureter).

2.4. Follow-Up Evaluation

Ultrasound examinations of abdominal cavities and a complete physical examination were performed 24, 72 h, and two weeks after the procedure to detect any postoperative complications. Additionally, information was obtained by telephone from animal owners about the health status of the operated animals at least one year after the procedure.

The resected kidney and fragments of the ureter from the two last operated cases (cats) were fixed in 10% formalin, and after obtaining histopathological preparations, they were also stained with hematoxylin and eosin (HE). The histopathological slides were observed under an Olympus BX53 microscope coupled with an Olympus UC90 camera (Olympus, Tokyo, Japan). For acquisition, the cellSens Standard V1 software was used (Olympus, Tokyo, Japan).

2.5. Statistical Analysis

The median, range, numbers, and percentages of the overall results were reported for the animals affected by the variables of interest. The available software (Microsoft Corp., Version 16.56, Microsoft, Albuquerque, NM, USA) was used for all the calculations.

3. Results

3.1. Clinical Results

Laparoscopic nephrectomies were performed on the three dogs and four cats that were referred to our clinic with moderate hydronephrosis and severe loss of renal parenchyma due to ureter obstruction. All animals that qualified for the procedure participated in observations for at least a year after the surgery. The median age at the time of laparoscopic nephrectomy was 6 years (range, 4–8 years) for dogs and 6.5 years (range, 4–8 years) for cats (Table 1). The body weight of the operated dogs was from 4 kg to 32 kg, (mean weight 13.2 kg) and was from 3 kg to 4 kg (mean weight 3.5 kg) for the cats. Based on the abdominal ultrasound examination made for reasons unrelated to urological problems, hydronephrosis was detected incidentally in each case presented in this paper. Preoperative urine culture results did not reveal any urinary tract infection. Preoperative morphological and biochemical parameter blood tests and urinalysis tests were within the reference range.

All procedures were performed completely laparoscopically without the need for conversion. The renal blood vessels (artery and vein) were closed in all operated animals using VSDs without the presence of intraoperative and postoperative bleeding. The ureter in the operated animals were also closed using VSDs, and the closure of the ureter lumen was evidenced by the lack of urine leakage from the ureter stump on the side of the urinary bladder.

Physical examination performed 24, 72 h, and two weeks after the procedure did not reveal any postoperative abnormalities. None of the dogs that underwent surgery experienced major early or late complications. Follow-up ultrasound examinations performed 24, 72 h, and two weeks after the procedure did not show any postoperative complications, such as the presence of free fluid in the abdominal cavity or dilatation of the remaining ureter stump. Control blood tests performed 72 h after the procedure did not reveal any abnormalities in morphological and biochemical parameters, which were within the reference range.

According to the information obtained from the owners, at least one year after the surgery, none of the animals had urinary system symptoms, including those related to the nephrectomy performed. In one cat, a stone was found in the contralateral kidney during a control ultrasound examination.

3.2. Histopathological Results

Histopathological evaluation of the resected ureters revealed closure of the lumen by coagulation of the walls on its end (Figure 2A,B—red arrows). Blurring to the structure of the ureter's muscular layer was observed. Initially, there were a few coagulated myocytes amongst the normal, unchanged muscular layer (Figure 2A,B—yellow arrows), as well as at the dissected end of the ureter where there were only coagulated myocytes visible (Figure 2A,B—red arrows). Here, only amorphous, highly basophilic band-like structures lying parallel to each other can be noted. Coagulated tissue did not bind together into one homogenous, amorphous compact mass but partially delaminated. A delicate fat tissue can be observed on the periphery. As is visible in Figure 2A, fragment of the epithelium before coagulation was not destructed (green arrow).

Histopathological evaluation was performed in two cats only. Owners of other patients did not consent for histopathological examination due to the economic reasons.

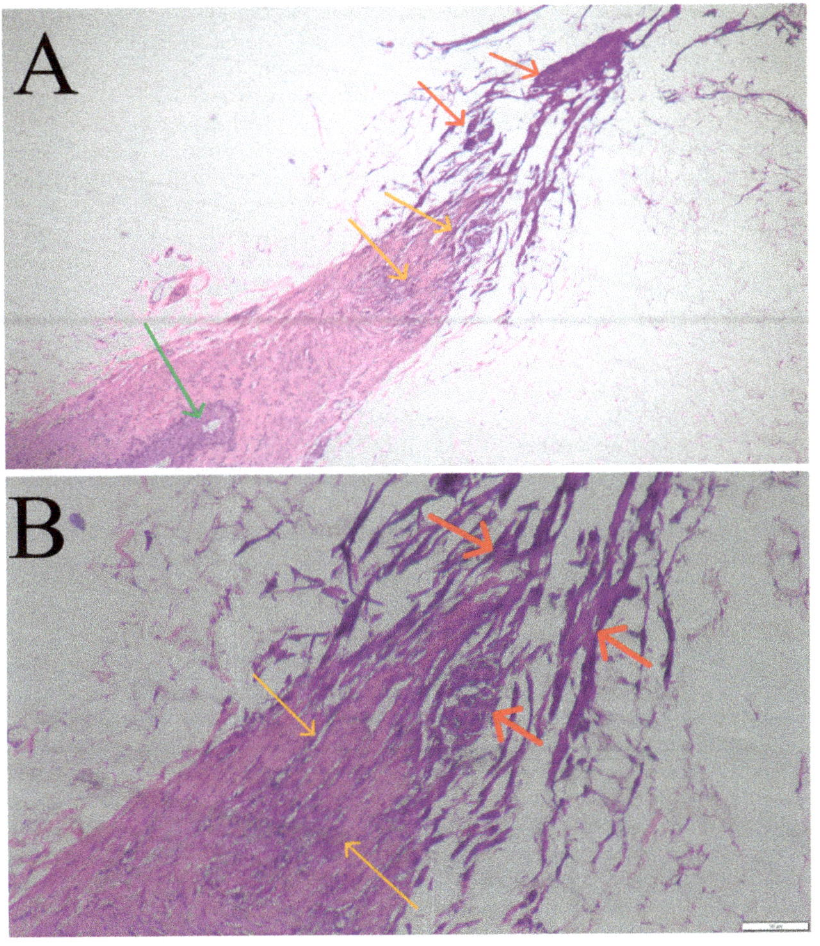

Figure 2. Histopathological images of the feline ureter occluded by vessel sealing device. (**A**)—Visible partial (among muscular layer—yellow arrows) and complete (at the dissected end of the ureter—red arrows) coagulation of the ureteral tissue and closure of its lumen. In the area of non-coagulated tissue, normal ureteral epithelium is visible (green arrow). Staining HE, magnification 40×. (**B**)—Visible partial (among muscular layer—yellow arrows) and complete (at the dissected end of the ureter—red arrows) coagulation of the ureteral tissue and closure of its lumen. Staining HE, magnification 100×.

4. Discussion

In this paper, we present the possibility of safe and effective use of VSDs in ureteral closure during nephrectomy in selected animals with moderate hydronephrosis without enlargement of the distal ureter. In the presented study, there were no complications such as uroabdomen or hemorrhage. Uroabdomen following nephrectomy due to proximal ureteral leak is the most serious postoperative complication immediately following hemorrhage from the renal blood vessels. Laparoscopic nephrectomy was feasible in 7/7 cases (100%) without the need for conversion. It should be noted that laparoscopic closure of the ureter by VSDs was carried out in patients without distension of the distal part of the ureter. Mayhew et al. [1] stated that, in some patients, e.g., in extensive cancers, conversion to an open approach was necessary. Conversion to an open approach may be necessary in cases where the affected ureter is severely dilated or adhesions to the surrounding structures can be observed, particularly in situations where it is impossible to visualize the ureterovesicular junction [1]. In this pilot study, the ureter was closed directly distally from the obstruction without dissecting the ureter at the uretrovesicular junction. Short-term postoperative follow-up abdominal ultrasound did not reveal hydroureter in the remaining part of the resected ureter in any patient that qualified for the procedure using VSDs. According to the authors' knowledge, the resecting of the ureter without its complete removal has not been described in veterinary literature before. The ureter was closed distally from the obstruction, where it was not dilated, and its diameter was corresponding to that in healthy animals. It should be noted that the biggest dog in this pilot study was 32 kg. Nephrectomy with complete ureter removal is traditionally recommended in small animals due to the fear of leaving a blind-ending ureteral stump, which may become a source of chronic infection and urine pooling [1]. Reviewing the literature, it can be noted that managing the distal ureter during laparoscopic nephro-ureterectomy is a recurring theme over the past decade [19,20]. Papers on what occurs to the ureteral stump post nephro-ureterectomy are limited [19,21]. Total removal of the ureter to the ureterovesicular junction in cases of urinary reflux is recommended in human pediatric medicine [22]. It should be noted that there is a possibility of complications related to the leaving of the ureteral stump post ne-phrectomy; however, there was no evidence within our patient group. The most common complication postoperatively during human surgery is infection of the remaining ureter including the ureteral wall, which shadows symptoms of pyelenophritis. There is the possibility that the infected ureteral stump may develop into an abscess. Suspicion of the ureteral stump abscess can be made if an oval cystic mass, adjacent to the bladder in ret-rovesical space, whose wall exhibits a contrast enhancement, which is found during computed tomography. Histopathological evaluation is required to confirm the diagnosis of empyema of the ureteral stump [22,23]. In post nephrectomy, a complication such as urinal reflux from the bladder requires the removal of the ureteral stump. A correlation between patients with high grade vesicoureteral reflux and complications which affected the stump were noted by Krarup and Wolf [24]. However, only 5% of cases exhibiting indications for stump removal were observed by Barroso et al. [25]. Moreover, ureteral stump removal was not necessary for any of the patients presenting with primary reflux into a single system. Therefore, there is a suggestion that the need for stumpectomy is linked to more complex cases. Advanced bipolar vessel sealing devices close vessels up to 7 mm in diameter with no risk of bleeding. In the vessels, the pressure is much higher than in the bladder; therefore, it is logical to use a VSD for closure of the ureter without the risk of urine leakage. Observations made within this pilot study confirmed this. Ureter closure was performed in two places adjacent to each other. First, the ureter was sealed with a VSD, then, after moving the device a few millimetres proximally, the ureter was sealed again and cut, allowing for kidney removal. Histopathological evaluation of resected ureters revealed their lumen sealed by coagulation. Observed changes in tissues resembled those caused by advanced bipolar coagulation within blood vessels [26]. The characteristic anatomical structure of the ureter opening into the bladder is an additional mechanism to prevent urine leakage. Ureters terminate at the cranial margin of the bladder neck after passing obliquely through its wall. The risk of urine reflux into ureters is reduced by this oblique intramural passage and intramural tension produced by pressure within the bladder [27]. The aforementioned mechanism probably does not work properly in cases of anatomical anomalies within the ureter and bladder connection and in patients after treatment of ectopic ureters. The authors of this pilot study decided to use a VSD for ureter closure when there were no signs of hydroureter in the sealing area. However, performing a complete ureter resection to the ureterovesical junction should be considered in cases of severe hydroureter [1]. Caution and further research are required in the use of VSDs for ureteral closure in patients with ureterovesicular junction abnormalities, or who had procedures for ectopic ureteral treatment performed in the past, as these cases can have an increased risk of urine reflux from the bladder to the ureters. This is an indication for removal of the whole ureter, up to the ureterovesicular junction. However, it requires longer laparoscopic dissection of the ureter, which sometimes can lead to additional complications, e.g., accidental damage of abdominal vessels or organs [28]. Another option is the intraoperative evaluation of stump leakage with retrograde pyelo-gram and/or cystogram. These procedures are used in human medicine for the evaluation of ureteral leakage [29]. In cases where postoperative free fluids within the abdomen are present, biochemical evaluation can be performed. A doubled increase in creatinine concentration in the fluid compared to the blood can suggest urine leakage [29].

Vessel sealing devices allow for the safe closure of blood vessels and thus reduce procedure times and the need to use other methods [1,11]. In the cases presented in this article, renal arteries and veins were closed using VSDs in all patients, the largest of whom was a 32 kg dog. Ligation of renal vessels can also be performed with laparoscopic hemo-clips, extra- and intracorporeal ligation, or en bloc stapling [1,30]. There are advantages to using VSD for the closure of renal vessels. The use of the device allows for a reduction in procedure time and surgery as a whole [11]. It should be noted that the significance of time reduction in laparoscopic nephrectomy with VSDs in veterinary medicine needs to be evaluated. It also reduces the need of clips, staplers, or ligation with suture [11–13,31]. Within the patients of this pilot study, there were no intra- and postoperative complications. Follow-up abdominal ultrasound and blood results did not show any abnormalities in the early postoperative period. Additional information collected from the owner after at least one year post-surgery, also did not show any long term problems or complications. One cat developed a stone in the remaining kidney eight months after the nephrectomy. However, the diagnosis was made by a referring doctor during control examination and the cat was asymptomatic.

The occurrence of no postoperative complications is probably the result of the criteria for selecting patients for the procedure. Kidney removal with a VSD was performed in patients with hydronephrosis and hydroureter where dilation of the ureter was within its proximal part, no more than half the length. Furthermore, the patients who qualified for the procedure were diagnosed with hydronephrosis accidentally and showed no clinical signs related to the urinary system inclusive of blood and urine results.

The authors are aware of some limitations of this pilot study. The group of patients was relatively small and there were strict criteria for qualifying for the procedure—the ureters were closed in places where tissue appeared macroscopically normal in appearance without visible distention. Closing of the ureters with VSDs was not performed on the hydroureters or at the level of the ureterovesicular junction. In addition, analysis of the uroliths causing obstruction was not performed. The lack of urolith analysis was a result of the individual owners' decisions despite the surgeon's recommendations.

5. Conclusions

The results of the present study suggest that ureteral closure using an advanced bipolar vessel sealing device is a safe and effective alternative for closing the unextended ureter during nephrectomy. With this technique in selected cases, there is no need to use vascular clips, staplers, or ligatures for ureteral closing. The procedure allows the performance of nephrectomy using only Kelly dissection forceps, an advanced bipolar vessel sealing device, and a monopolar hook.

Partial removal of the ureter is possible. Leaving the non-dilated distal part of the ureter in a properly working ureterovesicular junction appears to be safe and does not lead to the development of postoperative hydroureter complications in a relatively short time after surgery. This subject needs a longer term of observation and critical assessment of each operated case.

Author Contributions: Conceptualization, P.P.; methodology, P.P., B.L., K.S., A.A., Z.K., D.K.-N., S.D., A.S., and L.G.; formal analysis, P.P., L.G., A.A., Z.K., K.S., and B.L.; investigation, P.P.; resources, D.K.-N., A.S., and S.D.; data curation, P.P., S.D., and L.G.; writing—original draft preparation, P.P.; writing—review and editing, P.P., B.L., and L.G.; visualization, P.P. and S.D.; supervision, Z.K., A.A., S.D., B.L., and L.G. All authors have read and agreed to the published version of the manuscript.

Funding: This research received no external funding.

Institutional Review Board Statement: The animal study protocol was approved by the Institutional Review Board of Wroclaw University of Environmental and Life Sciences, Animal Welfare Advisory Team (19/2024; 4 November 2024).

Informed Consent Statement: Not applicable.

Data Availability Statement: Data is contained within the article.

Conflicts of Interest: The authors declare no conflicts of interest.

References

1. Mayhew, P.D.; Mehler, S.J.; Mayhew, K.N.; Steffey, M.A.; Culp, W.T.N. Experimental and Clinical Evaluation of Transperitoneal Laparoscopic Ureteronephrectomy in Dogs. *Vet. Surg.* **2013**, *42*, 565–571. [CrossRef]
2. Gookin, J.L.; Stone, E.A.; Spaulding, K.A.; Berry, C.R. Unilateral Nephrectomy in Dogs with Renal Disease: 30 Cases (1985–1994). *J. Am. Vet. Med. Assoc.* **1996**, *208*, 2020–2026. [CrossRef] [PubMed]
3. Rousset, N.; Abbondati, E.; Posch, B.; Owen, L.J.; Herrtage, M. Unilateral Hydronephrosis and Hydroureter Secondary to Ureteric Atresia, and Uterus Unicornis in a Young Terrier. *J. Small Anim. Pract.* **2011**, *52*, 441–444. [CrossRef]
4. Hart, E.; Singh, A.; Thomson, C.B.; Appleby, R.; Richardson, D.; Hocker, S.; Bernard, S.; Pinard, C. Outcomes after Transperitoneal Laparoscopic Ureteronephrectomy for the Treatment of Primary Renal Neoplasia in Seven Dogs. *Vet. Surg.* **2021**, *50* (Suppl. S1), O108–O115. [CrossRef] [PubMed]
5. Apostolopoulou, E.P.; Vlemmas, I.; Pardali, D.; Adamama-Moraitou, K.K.; Poutahidis, T.; Papadopoulou, P.L.; Brellou, G.D. Bilateral Renal Large B Cell Lymphoma in a Dog: A Case Report and Review of the Literature. *Vet. Sci.* **2021**, *8*, 258. [CrossRef] [PubMed]
6. Bennett, F. Unilateral Renal Cell Carcinoma in a Labrador Retriever. *Can. Vet. J.* **2004**, *45*, 860–862. [PubMed]
7. Ravizzini, P.I.; Shulsinger, D.; Guarnizo, E.; Pavlovich, C.P.; Marion, D.; Sosa, R.E. Hand-Assisted Laparoscopic Donor Nephrectomy versus Standard Laparoscopic Donor Nephrectomy: A Comparison Study in the Canine Model. *Tech. Urol.* **1999**, *5*, 174–178. [PubMed]
8. Marcovich, R.; Williams, A.L.; Seifman, B.D.; Wolf, J.S. A Canine Model to Assess the Biochemical Stress Response to Laparoscopic and Open Surgery. *J. Endourol.* **2001**, *15*, 1005–1008. [CrossRef]
9. Yoder, B.; Wolf, J.S. Canine Model of Surgical Stress Response Comparing Standard Laparoscopic, Microlaparoscopic, and Hand-Assisted Laparoscopic Nephrectomy. *Urology* **2005**, *65*, 600–603. [CrossRef] [PubMed]
10. Kim, Y.K.; Park, S.J.; Lee, S.Y.; Suh, E.H.; Lee, L.; Lee, H.C.; Yeon, S.C. Laparoscopic Nephrectomy in Dogs: An Initial Experience of 16 Experimental Procedures. *Vet. J.* **2013**, *198*, 513–517. [CrossRef] [PubMed]
11. Constant, D.L.; Florman, S.S.; Mendez, F.; Thomas, R.; Slakey, D.P. Use of the LigaSure Vessel Sealing Device in Laparoscopic Living-Donor Nephrectomy. *Transplantation* **2004**, *78*, 1661–1664. [CrossRef] [PubMed]
12. Park, K.S.; Sim, Y.J.; Jung, H. Migration of a Hem-o-Lok Clip to the Ureter Following Laparoscopic Partial Nephrectomy Presenting With Lower Urinary Tract Symptoms. *Int. Neurourol. J.* **2013**, *17*, 90–92. [CrossRef] [PubMed]
13. Kurz, D.A.; Mucksavage, P. The Endoscopic Removal of Eroded Foreign Bodies in the Ureter. *J. Endourol. Case Rep.* **2016**, *2*, 30. [CrossRef] [PubMed]
14. Moser, R.L.; Narepalem, N. Erosion of Hem-o-Lok Clips at the Bladder Neck after Robot-Assisted Radical Prostatectomy. *J. Endourol.* **2009**, *23*, 949–951. [CrossRef]
15. Gardeweg, S.; Bockstahler, B.; Duprè, G. Effect of Multiple Use and Sterilization on Sealing Performance of Bipolar Vessel Sealing Devices. *PLoS ONE* **2019**, *14*, e0221488. [CrossRef] [PubMed]
16. Harold, K.L.; Pollinger, H.; Matthews, B.D.; Kercher, K.W.; Sing, R.F.; Heniford, B.T. Comparison of Ultrasonic Energy, Bipolar Thermal Energy, and Vascular Clips for the Hemostasis of Small-, Medium-, and Large-Sized Arteries. *Surg. Endosc.* **2003**, *17*, 1228–1230. [CrossRef] [PubMed]
17. Mayhew, P.D.; Singh, A. Laparoscopic Ureteronephrectomy. In *Small Animal Laparoscopy and Thoracoscopy*; John Wiley & Sons, Ltd.: Hoboken, NJ, USA, 2022; pp. 233–240, ISBN 978-1-119-66691-2.
18. Fossum, T.W. *Small Animal Surgery Textbook—E-Book: Small Animal Surgery Textbook—E-Book*; Elsevier Health Sciences: Amsterdam, The Netherlands, 2012; ISBN 978-0-323-17121-2.
19. Androulakakis, P.A.; Stephanidis, A.; Antoniou, A.; Christophoridis, C. Outcome of the Distal Ureteric Stump after (Hemi) Nephrectomy and Subtotal Ureterectomy for Reflux or Obstruction. *BJU Int.* **2001**, *88*, 586–589. [CrossRef]
20. Arora, S.; Yadav, P.; Ansari, M.S. Diagnosis and Management of Symptomatic Residual Ureteral Stump after Nephrectomy. *Case Rep.* **2015**, *2015*, bcr2015209441. [CrossRef] [PubMed]
21. Cain, M.P.; Pope, J.C.; Casale, A.J.; Adams, M.C.; Keating, M.A.; Rink, R.C. Natural History of Refluxing Distal Ureteral Stumps after Nephrectomy and Partial Ureterectomy for Vesicoureteral Reflux. *J. Urol.* **1998**, *160*, 1026–1027. [CrossRef]
22. Escolino, M.; Farina, A.; Turrà, F.; Cerulo, M.; Esposito, R.; Savanelli, A.; Settimi, A.; Esposito, C. Evaluation and Outcome of the Distal Ureteral Stump after Nephro-Ureterectomy in Children. A Comparison between Laparoscopy and Retroperitoneoscopy. *J. Pediatr. Urol.* **2016**, *12*, 119.e1–119.e8. [CrossRef] [PubMed]

23. Labanaris, A.P.; Zugor, V.; Smiszek, R.; Nützel, R.; Kühn, R. Empyema of the Ureteral Stump. An Unusual Complication Following Nephrectomy. *Sci. World J.* **2010**, *10*, 380–383. [CrossRef] [PubMed]
24. Krarup, T.; Wolf, H. Refluxing Ureteral Stump. *Scand. J. Urol. Nephrol.* **1978**, *12*, 181–183. [CrossRef] [PubMed]
25. Barroso, U.; Calado, A.A.; Filho, M.Z. The Role of Refluxing Distal Ureteral Stumps after Nephrectomy. *J. Pediatr. Surg.* **2002**, *37*, 653–656. [CrossRef] [PubMed]
26. Okhunov, Z.; Yoon, R.; Lusch, A.; Spradling, K.; Suarez, M.; Kaler, K.S.; Patel, R.; Hwang, C.; Osann, K.; Huang, J.; et al. Evaluation and Comparison of Contemporary Energy-Based Surgical Vessel Sealing Devices. *J. Endourol.* **2018**, *32*, 329–337. [CrossRef]
27. Adin, C. *Nephrology and Urology of Small Animals*; John Wiley & Sons: Hoboken, NJ, USA, 2014; pp. 306–328, ISBN 978-0-8138-1717-0.
28. Prządka, P.; Liszka, B.; Skrzypczak, P.; Kubiak-Nowak, D.; Borawski, W.; Juźwiak, Ł.; Kiełbowicz, Z.; Patkowski, D. Laparoscopic Assisted Percutaneous Herniorrhapy in Dogs Using PIRS Technique. *PLoS ONE* **2020**, *15*, e0235899. [CrossRef]
29. Rosenfeld, J.; Boehm, D.; Raikar, A.; Coskey, D.; Lee, M.; Ji, E.; Lee, Z. A Review of Complications after Ureteral Reconstruction. *Asian J. Urol.* **2024**, *11*, 348–356. [CrossRef] [PubMed]
30. Schatloff, O.; Lindner, U.; Lindner, A. Current Status of En Bloc Stapling of the Renal Hilum during Laparoscopic Nephrectomy. *J. Laparoendosc. Adv. Surg. Tech. A* **2010**, *20*, 631–633. [CrossRef] [PubMed]
31. Cormio, L.; Massenio, P.; Lucarelli, G.; Di Fino, G.; Selvaggio, O.; Micali, S.; Carrieri, G. Hem-o-Lok Clip: A Neglected Cause of Severe Bladder Neck Contracture and Consequent Urinary Incontinence after Robot-Assisted Laparoscopic Radical Prostatectomy. *BMC Urol.* **2014**, *14*, 21. [CrossRef] [PubMed]

life

MDPI

Article

Thermographic Scan of the Thoracolumbar Area in Dogs with Acute Intervertebral Disc Extrusion (IVDE): A Retrospective Study

Cristian Zaha [1], Liliana Cărpinișan [1,*], Larisa Schuszler [1], Nistor Paula [1], Tudor Căsălean [1], Tiana Florea [2], Văduva Cristina [3], Bogdan Sicoe [4], Ciprian Rujescu [5] and Roxana Dascălu [1]

1 Surgery Clinic, Faculty of Veterinary Medicine, University of Life Sciences "King Michael I", 300645 Timisoara, Romania; cristian.zaha@usvt.ro (C.Z.); larisaschuszler@usvt.ro (L.S.); paula.nistor@usvt.ro (N.P.); Tudor-Mihai.Casalean.FMV@usvt.ro (T.C.); roxanadascalu@usvt.ro (R.D.)
2 Dermatology Department, Faculty of Veterinary Medicine, University of Life Sciences "King Mihai I", 300645 Timisoara, Romania; tijana.florea@usvt.ro
3 Internal Medicine, Faculty of Veterinary Medicine, University of Life Sciences "King Michael I", 300645 Timisoara, Romania; cristina.vaduva@usvt.ro
4 Diagnostic Imaging, Faculty of Veterinary Medicine, University of Life Sciences "King Michael I", 300645 Timisoara, Romania; bogdan.sicoe@usvt.ro
5 Management and Rural Development Department, Faculty of Management and Rural Tourism, University of Life Sciences "King Michael I", 300645 Timisoara, Romania; rujescu@usvt.ro
* Correspondence: lilianacarpinisan@usvt.ro

Abstract: Background: several authors have documented variations in local temperature in both horses and dogs presenting acute intervertebral disc extrusion (IVDE) along the entire spinal column. However, none have demonstrated distinct temperature differences between healthy animals and those with IVDE. A retrospective study was conducted to assess the efficacy of thermography at evaluating local temperature and thermal patterns in healthy dogs as well in those with IVDE across the T11–L3 area. Methods: the study included 20 healthy dogs and 32 dogs with IVDE. For both groups of dogs, the thoracolumbar region was trimmed and, subsequently, scanned using the Flir E50 thermography device. The Flir Tool software was used to analyze three designated areas (Bx1, Bx2, Bx3) within the thoracolumbar region by comparing the average temperature of the minimum, maximum, and mean temperature recordings between the two groups. Results: the thermal pattern and the local temperature of the thoracolumbar area present differences between healthy dogs and those with IVDE. Conclusions: we recommend thermographic scanning of the thoracolumbar area to find differences in local temperature between healthy dogs and those with intervertebral disc extrusion. Further investigations are required to differentiate between disc extrusion that exhibits lateralization to the right or left.

Keywords: dogs; intervertebral disc extrusion; thermography

1. Introduction

Intervertebral disc degeneration is a common disease and one of the most serious problems worldwide in chondrodystrophic breeds [1]. Intervertebral disc disease (IVDD) in the thoracolumbar region is the leading cause of paraplegia in chondrodystrophic dogs, commonly occurring between 3 and 7 years of age [1–3]. Several authors have identified a higher frequency of disc extrusion in the T11–L3 region in dogs, as this is the most mobile area during running. This increased mobility contributes to heightened mechanical stress on the intervertebral discs, ultimately culminating in their progressive degeneration [4]. The likelihood of successful recovery is closely tied to the speed of diagnosis and treatment following disc material extrusion. Intervertebral disc injuries are caused by aging and mechanical stress, both of which contribute to degenerative changes in the spine [1,5]. Intervertebral disc disease is divided into acute intervertebral disc extrusion (IVDE) and intervertebral disc protrusion [3,6]. Also, a genetic factor is involved in IVDE, as independent genome-wide association studies for skeletal dysplasia and IVDE have identified a highly expressed *FGF4* retrogene on CFA12 which is associated with both IVDE and chondrodystrophy in dogs [7].

The clinical signs of IVDD in dogs depend on the type and location of the disc extrusion and may range from pain and discomfort to severe neurological deficits [8,9]. Types of clinical signs vary from spinal hyperesthesia, ataxia, paraparesis, paraplegia with or without deep pain perception, and urinary and fecal incontinence [1,9].

Various costly methods are available to diagnose intervertebral disc herniation, and most require general anesthesia to obtain clear imaging. Radiography (Rx), myelography, computed tomography (CT), and magnetic resonance imaging (MRI) are used for diagnosis of the intervertebral disk [10–12]. This obtained images can be conventional or may require the administration of the contrast solution in the subarachnoidal space in case of IVDE [13,14]. CT provides more precise information and is faster than myelography, especially in chondrodystrophic breeds. A study involving 20 dogs compared the accuracy of CT and myelography in diagnosing acute intervertebral disc disease [13]. The findings revealed that CT was 90% accurate, while myelography was 88% accurate in identifying the primary location of disc herniation [13,15]. CT also correctly predicted the lateralization of disc material in 96% of the cases, compared to 92% for myelography [11]. MRI, with an accuracy of 98.5% in detecting IVDE [16] and also providing transverse imaging, outperforms CT when the disc material is not mineralized, and, in such cases, a contrast medium may be needed [11,17]. However, using CT with a subarachnoid contrast medium negates the advantage of diagnosing disc herniation without the side effects associated with myelography [18].

Medical infrared imaging, also known as infrared thermography (IRT), is a non-invasive method that captures infrared radiation emitted from the body's surface, creating a visual representation of body temperature. It can also identify abnormal physiological changes in both humans and animals [19–21]. The diagnostic accuracy of infrared thermography (IRT) for detecting spinal disease in humans is 80%, while it demonstrates a 90% accuracy in distinguishing normal dogs from those affected by IVDE [21–23].

IRT can be a useful tool for the evaluation of body surface temperature changes in response to exercise [19,20], for diagnosing diseases affecting the canine locomotor system [24,25], and for assessing the response of an organism's body to drug administration [23]. The highest surface temperatures of the body have been observed in the trunk and lumbosacral area, with a temperature gradient across the limbs, where proximal regions are warmer than the distal areas [26].

In human medicine, the normal thermogram of a human spine has been characterized by a central zone of decreased heat emission in the region of the spinal processes from the cervical to the lower lumbosacral spine [27,28]. Pathological alterations in the thermal pattern include focal changes in temperature over the site of an active lesion as well asymmetrical hyperthermia and hypothermia distal to the lesions [29].

Turner T. reported the identification of a localized region of hyperthermia adjacent to the affected disc in horses exhibiting neurological injuries [30].

Grossbard et al. aimed to assess the efficacy of infrared thermography (IRT) in identifying dogs with intervertebral disc disease (IVDD) and evaluate whether the normal thermal pattern is restored 10 weeks following decompression surgery [21].

Given that an increase in local tissue temperature typically follows an inflammatory process, the objective of this study is to compare local temperature measurements and evaluate thermal patterns between healthy dogs and those affected by IVDE.

We hypothesize that the local temperature and thermal pattern of the thoracolumbar area will exhibit significant changes, with an increase in local temperature and a modified thermal pattern in dogs with IVDE.

2. Materials and Methods

A retrospective study was performed at the Surgery Clinic of the Faculty of Veterinary Medicine of Timisoara between March 2021 and October 2024.

2.1. Animal Selection and Clinical Examination

During this period, 75 dogs from chondrodystrophic breeds (French bulldog, dachshund, bichon frise, Lhasa apso, shih tzu, chihuahua, cocker spaniel, Havanese, and mixed breeds), aged between 3 years and 2 months and 7 years and 4 months and weighing between 9 kg and 15 kg, presented with an acute form of paraplegia. The inclusion criteria for the study were as follows: localization of disc extrusion in the T11–L3 region, presentation for clinical examination within 12 h of paraplegia onset, paresis with decreased and absent proprioception, paralysis with deep pain perception present, absence of urinary or fecal incontinence, and rectal temperature between 38 and 39.5 °C. The exclusion criteria were as follows: weight exceeding 15 kg, presence of disc extrusion outside the T11–L3 area, presence of two or more disc extrusions in the examined region, spondylosis in the T11–L3 or L7–S1 segments, discospondylitis administration of anti-inflammatory medication at another clinic or private practice, urinary or fecal incontinence, presence of hemivertebrae, tumors, or thromboembolism.

A clinical and neurological examination was performed by two veterinarians (Z.C. and D.R.), both with clinical experience in musculoskeletal disorders, who were not blinded to each other's results. The neurological examination included testing proprioceptive deficits in the hind limbs, cutaneous trunchi reflex, patellar reflex, anal reflex, and superficial and deep pain sensation [3]. Following the neurological examination, each dog was assigned a neurological score based on a modified version of the scale published by Wheeler and Sharp [4].

A control group, consisting of 30 dogs from chondrodystrophic breeds (French bulldog, dachshund, shih tzu), underwent clinical and neurological examinations performed by the same veterinarians (Z.C. and D.R.). The inclusion criteria for the control group required the absence of anti-inflammatory drug administration, no reported history of lameness within the preceding month, no radiographic changes to the spinal vertebrae, and a rectal temperature between 38 and 39.5 °C.

2.2. Thermal Imaging and Data Recording

Following the clinical examination, the hair in the dorsal region of the animal, covering the T7–L7 segment, was removed using a groomer clipper, both for the dogs in the study group and for those in the control group.

Following hair removal, the animals were allowed a 30 min acclimatization period under controlled environmental conditions, specifically an air temperature of 20–21 °C, humidity between 70% and 75%, and no air currents within the room. Thermographic imaging was performed by the same individual on both the dogs with disc extrusions and those in the control group. The operator was positioned 1 m from the dog at a 90° angle to the thoracolumbar region. Thermographic images (Figure 1a,b) were obtained using the FLIR E50 thermography device (FLIR Systems Inc., Wilsonville, OR, USA), with the following parameters: 0.95 emissivity and a resolution of 240 × 180 for each thermographic image. The temperature range was set from −20 °C to 650 °C, with a sensitivity of ≤0.05 °C.

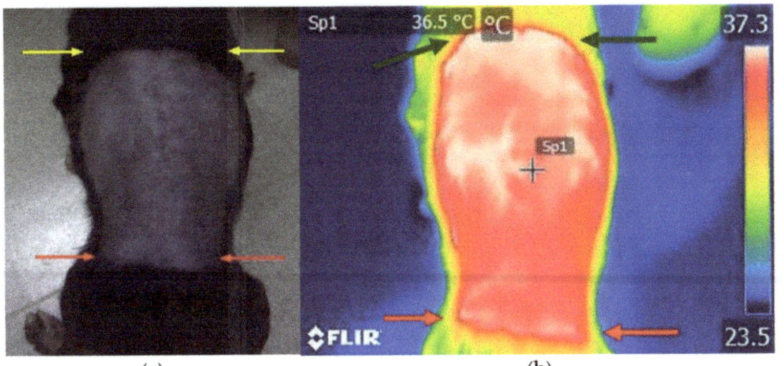

(a) (b)

Figure 1. Photo of the thoracolumbar area in a paraplegic dog: (**a**) normal image from the FLIR E50 camera, (**b**) thermographic image from the FLIR E50 camera. Arrows: (**a**) yellow arrows—T7 level, red arrows—L7 level; (**b**) green arrows—T7 level, red arrows—L7 level.

The obtained thermographic images were processed and interpreted by the same operator, using the FLIR Tools software 5.X for image analysis. For the interpretation of the thermographic images, a rectangle (Bx1, Bx2, and Bx3) with dimensions of 94 × 42 pixels was drawn.

For the dogs in the control group, thermographic images were obtained between 9 a.m. and 2 p.m., while, for the dogs in the study group, thermographic images were obtained between 9 a.m. and 4 p.m., depending on their presentation for the clinical examination.

2.3. Imagistic Investigations

The diagnosis of the spinal cord compression was established after a radiological exam and a CT scan. The radiological exam required two views (latero-lateral and ventro-dorsal) using the Siemens Multix Swing device. For the CT scan, the animal required anesthesia by administering to the animal a combination of medetomidine in doses of 0.02 mg/kg (Domitor 2%, Montero, Romania) and ketamine in doses of 5 mg/kg (Ketamidor 100 mg/mL, Richter Pharma, Austria), and, if necessary, it was supplemented with propofol in doses of 4 mg/kg (Propofol 10 mg/mL, Braun, Romania).

2.4. Statistical Analysis

Two groups of dogs were subjected to a comparative analysis of the values obtained through thermographic scanning, which included a control group (n = 20) and a study group (n = 32). For each case, three rectangular regions centered on the thoracolumbar area, labeled Bx1, Bx2, and Bx3, were examined. This resulted in a set of statistical data derived from the observed temperatures, which were expressed in degrees Celsius. For each thermographic image, the mean of the maximum, minimum, and average temperature values recorded within the designated rectangle in the FLIR Tools software was considered. The comparisons involved examining the differences between the study group and the control group, based on the location from which the data were obtained: Bx1 first, then Bx2, and, separately, Bx3. Additionally, the means of the minimum, maximum, and average values were compared separately. This resulted in nine comparisons for 18 datasets. A two-sample t-test was performed using SAS Studio, following normality testing of the distributions with the Shapiro–Wilk test.

3. Results

3.1. Animal Information

Out of 75 consecutively examined cases, including clinical and neurological exams, 47 cases were selected for radiographic imaging (27 males and 20 females). After the exclusion criteria were applied to the examined dogs, twenty-eight of them were excluded for the following reasons: four dogs weighed over 15 kg, six dogs had chronic paralysis, 10 dogs presented without deep pain perception, and eight dogs presented with urinary or fecal incontinence.

Following the clinical and neurological examinations, five dogs were classified with grade 1 deficits, 19 dogs with grade 2 deficits, 14 dogs with grade 3 deficits, eight dogs with grade 4 deficits, and one dog with grade 5 deficits. All results were obtained using the scale of Wheeler and Sharp [4].

Following radiographic imaging, 39 cases were selected for computed tomography (CT). A total of eight dogs were excluded from the study due to the following reasons: spondylosis in the L7–S1 region (two dogs), T13–L1 region (three dogs), L1–L2 region (one dog), and L2–L3 region (two dogs). The dogs that met the inclusion criteria were aged between 3 and 8 years (mean age: 4.8 years), with a mean weight of 8.55 kg (range: 4.32–14.80 kg). The study included the following breeds: six French bulldogs, 10 dachshunds, seven bichon frises, two Lhasa apsos, six shih tzus, two chihuahuas, two cocker spaniels, one Havanese, and three mixed breeds.

Out of the 30 dogs in the control group, 20 dogs were included in the study. The excluded dogs did not meet the following criteria: four dogs exhibited pain upon palpation in the dorsal lumbar region, two dogs showed spondylosis in the T13–L1 and L1–L2 regions, and four other dogs presented spondylosis in the L7–S1 region. The dogs in the control group belonged to the following breeds: shih tzu (five dogs), dachshunds (eight dogs), bichon frise (four dogs), cocker spaniels (three dogs). They were aged between 4 and 7 years (mean age: 5.2 years), with a mean weight of 9.35 kg (range: 5.4–14.2 kg).

The rectal temperature of all dogs was between 38 and 39.5 °C and was measured at the start of the examination.

3.2. Imagistic Investigation

A total of 52 conventional radiographs were performed for the study group (n = 32) and the control group (n = 20). Both ventro-dorsal and latero-lateral views were taken. Following the radiographs, lesions that could be associated with disc extrusion were identified, such as the narrowing of the intervertebral space in seven dogs and increased opacity of the intervertebral foramen in three dogs.

A total of 39 computed tomography scans were selected, and the results identified the following locations: four in the T11–T12 space, six in the T12–T13 space, 12 in the T13–L1 space (Figure 2a,b), six in the L1–L2 space, and four in the L2–L3 space. In nine cases, the disc material was located beneath the spinal cord, in 12 cases the disc material was identified with left lateralization, and, in 11 cases, the disc material was located on the right side. For seven conventional computed tomography scans, no disc extrusions were identified, and these cases were, therefore, excluded.

A total of 32 cases were selected following neurological, radiographic, and computed tomography examinations for the thermographic imaging study.

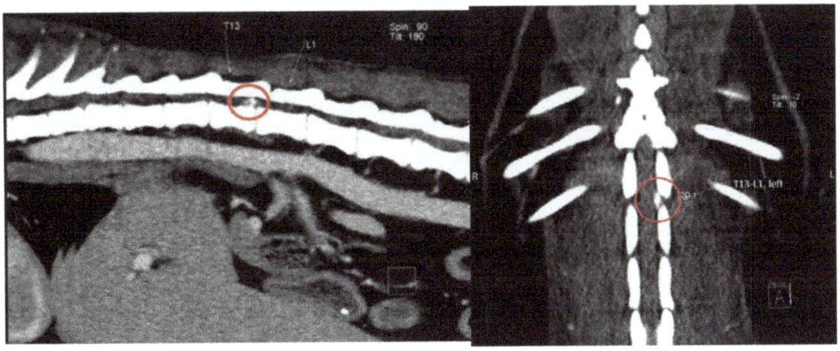

(a) (b)

Figure 2. Computed tomography image from a dog with T13–L1 disc extrusion: (**a**) sagitally reconstructed view of the vertebral column, T13—13th thoracic vertebra, L1—1th lumbar vertebra, (**b**) dorsally reconstructed view of the vertebral column, T13—13th thoracic vertebra, L1—1th lumbar vertebra, red circle—extruded discal material.

3.3. Thermography Scan

3.3.1. Thermography Scan of the Control Group

Different color variations were observed in the clipped area between T7–L7. There were observed alternations in local temperature, with warmer or cooler areas in all dogs of the control group (Figure 3a,b).

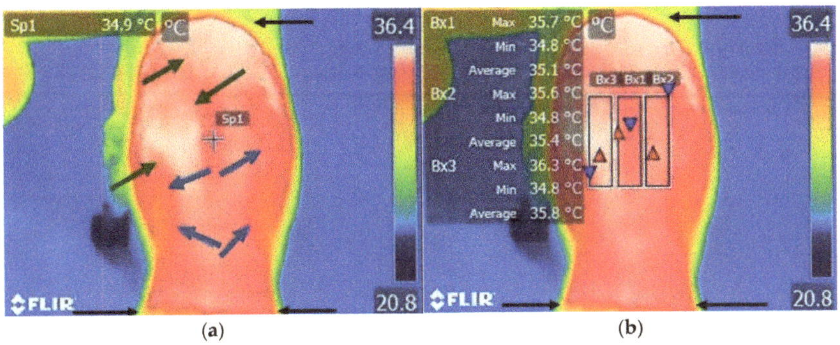

(a) (b)

Figure 3. Thermographic image of the thoraco-lumbar area of a healthy dog. (**a**) Scanning image without FLIR Tools software analysis: black arrow—clipped hair area between the T7 and L7 level, green arrow—variation areas of increase in temperature, blue arrow—variation areas of low temperature, spot—local temperature of the skin. (**b**) Scanning image with FLIR Tools software analysis: black arrow—clipped hair area between the T7 and L7 level, Bx1—area of interest centered on the vertebrae, Bx2—area of interest on the right side of the vertebrae, Bx3—aria of interest on the left side of the vertebrae; red triangle spot—maximum temperature recorded in the interest area; blue triangle spot—minimum temperature recorded in the interest area.

3.3.2. Thermography Scan of the Study Group

A high intensity temperature area was observed in the T11–L3 area in all dogs from the study group. Areas with cooler temperatures were observed at the margin of the T11–L3 area (Figure 4a,b).

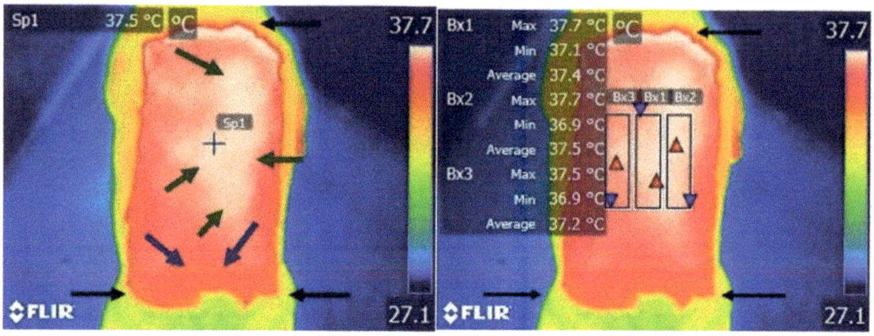

(**a**) (**b**)

Figure 4. Thermographic image of the thoraco-lumbar area in a dog with T13–L1 disc extrusion. (**a**) Scanning image without FLIR Tools software analysis: black arrow—clipped hair area between the T7 and L7 level, green arrow—area of increased temperature, blue arrow—area of lower temperature, spot—local temperature of the skin. (**b**) Scanning image with FLIR Tools software analysis: black arrow—clipped hair area between the T7 and L7 level, Bx1—area of interest centered on the vertebrae, Bx2—area of interest on the right side of the vertebrae, Bx3—area of interest on the left side of the vertebrae; red triangle spot—maximum temperature recorded in the interest area; blue triangle spot—minimum temperature recorded in the interest area.

3.4. Group Comparison

3.4.1. Comparisons Within the Bx1 Region

Regarding the comparisons in Bx1, the series consisting of the minimum values for the control group resulted in an average reading of 35.17 °C. The series consisting of the minimum values for the study group yielded an average result of 36.9 °C. The differences were statistically significant, with a t-value of −15.29 and $p < 0.001$. This indicates that, with respect to these values, animals in the study group exhibited higher local thermal values compared to those in the control group (Figure 5).

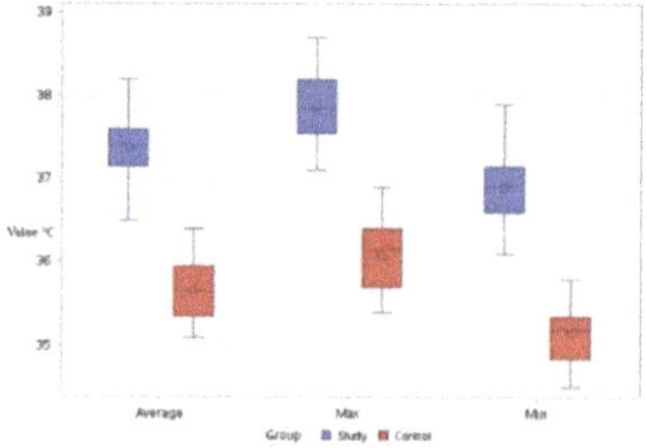

Figure 5. Comparative boxplot referring to the results obtained for Bx1 between the study group (blue mark) and the control group (red mark); distribution of the average, maximum, and minimum temperatures; o—outliers.

Comparing the two groups regarding the maximum temperatures observed in the Bx1 region, an average temperature of 36.08 °C was determined for the control group and an average temperature of 37.87 °C was determined for the study group. In this case as well, the local temperatures in the study group were higher than those in the control group. The differences were statistically significant, with a t-value of −15.46 and $p < 0.001$ (Figure 5).

Comparing the two groups with respect to the average temperature recorded in the Bx1 region, an average of 35.64 °C was observed in the control group and an average of 37.38 °C was observed in the study group. In this case as well, the local temperatures in the study group were higher than those in the control group, with statistically significant differences, t = −18.90, $p < 0.001$ (Figure 5).

3.4.2. Comparisons Within the Bx2 Region

Regarding the comparisons in Bx2, the series consisting of the minimum values for the control group resulted in an average reading of 34.71 °C. The series consisting of the minimum values for the study group yielded an average result of 36.75 °C. The differences were statistically significant, with a t-value of −15.70 and $p < 0.001$. This indicates that, with respect to these values, animals in the study group exhibited higher local thermal values compared to those in the control group (Figure 6).

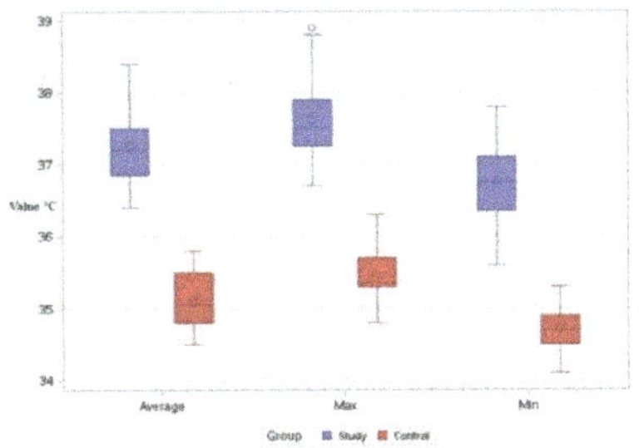

Figure 6. Comparative boxplot referring to the results obtained for Bx2 between the study group (blue mark) and the control group (red mark); distribution of the average, maximum, and minimum temperatures; o—outliers.

Comparing the two groups with respect to the maximum temperatures observed in the Bx2 region, an average of 35.45 °C was determined for the control group and an average of 37.61 °C was determined for the study group. In this case as well, the local temperatures in the study group were higher than those in the control group. The differences were statistically significant, with a t-value of −15.39 and $p < 0.001$ (Figure 6).

Comparing the two groups with respect to the average temperature recorded in the Bx2 region, an average of 35.10 °C was observed in the control group and an average of 37.25 °C was observed in the study group. In this case as well, the local temperatures in the study group were higher than those in the control group, with statistically significant differences, t = −15.76, $p < 0.001$ (Figure 6).

3.4.3. Comparisons Within the Bx3 Region

Regarding the comparisons in Bx3, the series consisting of the minimum values for the control group resulted in an average reading of 34.79 °C, while the minimum values for the study group yielded an average result of 36.71 °C. The differences were statistically significant, with a t-value of −13.51 and $p < 0.001$. This indicates that, with respect to these values, animals in the study group exhibited higher local thermal values compared to those in the control group (Figure 7).

Comparing the two groups with respect to the maximum temperatures observed in the Bx3 region, an average of 35.56 °C was determined for the control group and an average of 37.59 °C was determined for the study group. In this case as well, the local temperatures in the study group were higher than those in the control group. The differences were statistically significant, with a t-value of −13.54 and $p < 0.001$ (Figure 7).

Comparing the two groups with respect to the average temperature recorded in the Bx3 region, an average of 35.23 °C was observed in the control group and an average of 37.24 °C was observed in the study group. In this case as well, the local temperatures in the study group were higher than those in the control group, with statistically significant differences, $t = -14.09$, $p < 0.001$ (Figure 7).

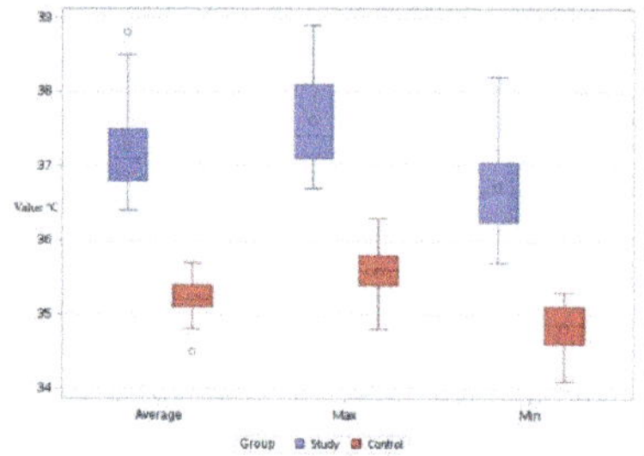

Figure 7. Comparative boxplot referring to the results obtained for Bx3 between the study group (blue mark) and the control group (red mark); distribution of the average, maximum, and minimum temperatures; o—outliers.

4. Discussion

Thermographic scanning of the thoraco-lumbar region provides important data to differentiate dogs with IVDE from healthy dogs; an aspect that is highlighted is the confirmation of the hypothesis regarding the differences among the minimum, maximum, and mean temperatures considered from the Bx1, Bx2, and Bx3 areas, which were identified in both groups of dogs. The results identified after scanning the healthy dogs and those with IVDE revealed that the values of temperature from the selected areas and the thermal patterns obtained presented changes, with increased temperatures in IVDE dogs.

Dogs with acute intervertebral disc extrusions presented a larger area of increased temperature in the thoraco-lumbar region compared with the same thermal pattern analyzed in healthy dogs. Furthermore, the temperature values recorded in the region of interest (T11–L3) were elevated by more than 1 °C in the study group when compared to the control group. Based on a study performed by Lahiri et al. [31], a local temperature that is 0.7–1.0 higher compared with the surrounding tissue is attributed to an inflammation.

A study conducted by Turner on horses with neurological injuries identified an area of hyperthermia at the site of the injury [30]. Similar findings, including an increase in local skin temperature, have been observed in humans with cervical intervertebral disc disease (IVD), providing valuable data not only for localizing disc herniation but also for detecting its presence [31]. It is crucial to differentiate between temperature increases associated with IVDE and those resulting from other pathological conditions, as thermographic scans alone cannot identify the underlying cause of changes in the thermal pattern, such as tumors or inflammation [3].

Intervertebral disc herniation is classified into Hansen type I and Hansen type II categories [3,32]. Hansen type I herniation predominantly occurs in chondrodystrophic breeds, such as dachshunds, and is typically associated with more severe clinical manifestations compared to Hansen type II [33]. In Hansen type I, there is often acute or subacute extrusions of the nucleus pulposus from a chondroid disc, represented by more severe neurological symptoms [1,7,14]. In our study, the majority of the dogs affected by disc extrusion Hansen type I where from the dachshund breed and presented clinical signs such as acute paralysis and thoraco-lumbar pain. The clinical signs of IVD vary from spinal hyperesthesia to paraplegia with or without proprioception and paralysis with or without deep pain perception [1]. In our study, the most frequent clinical sign was represented by paralysis with deep pain perception. In a study performed by Coates et al., more than 53% of the cases were dachshunds [1].

In our study, we designated the region T11–L3 as the area of interest, as this region is the most affected by disc extrusion. Wheeler and Sharp [4] noticed that, in relation to thoraco-lumbar disc disease, over 50% of the lesions were found at the T12–T13 and T13–L1 discs, while more than 85% occurred between the T11–T12 and L2–L3 levels [32]. The most common site affected by disc extrusion in our study was the T13–L1; a similar observation was made also by other authors in their study [14].

Radiographic evaluation allowed us to identify changes in 10 dogs, e.g., the narrowing of the intervertebral space and the increased opacity in the vertebral foramen; similar observations have been made by others authors in their studies [1,32]. Compared with radiography, CT scans can acquire a diagnostic index without an invasive procedure, require a short imagining time compared with nuclear magnetic resonance imaging, and enable a detailed visualization of the extruded disc material compared with radiography examination [6,10]. In our study, thirty-two disc extrusions were identified through conventional CT, but, in seven cases, conventional CT did not detect the extrusion of the disc material. Some authors have noted that myelography is necessary in situations where the nucleus pulposus does not degenerate or mineralize before extruding and compressing the spinal cord [14,27,33]. Conversely, when the extruded material is not mineralized and no other signs of disc extrusion are present, like the vacuum phenomenon, vertebral endplate sclerosis, and disc space narrowing, a myelography or CT myelography is required for accurate diagnosis and proper localization [6,34].

During the thermographic measurements, the ambient temperature in the examination room was maintained at 21 °C. Each dog, both in the control group and in the study group, underwent an acclimatization period of 30 min. The thoraco-lumbar region was not touched by the operator for at least 30 min prior to the scan. During handling, one hand was placed on the abdominal region and the other on the inguinal region. A temperature of 21 °C has also been cited by other authors as not influencing local skin temperature or thermoregulation in animals [35–37]. To prevent artifacts in the thoraco-lumbar area, other authors have also employed similar handling techniques to avoid contact with the area of interest [24,38,39].

Presence of the subcutaneous fat tissue in the L3–L5 vertebral area could have an insulating effect that reduces heat loss through the skin, appearing on thermographic images as an alternation in local temperature [40].

Numerous studies have been conducted to explore the impact of individual animal characteristics on IRT outcomes in healthy animals. Temperature differences between adjacent haired and non-haired skin indicate that the hair absorbs a portion of the radiated heat, thereby preventing this energy from being detected by an infrared camera [41]. Skin color can influence IRT readings, as black areas are typically warmer than adjacent white areas. In zebras, for example, black stripes are warmer than the white stripes during the day, likely due to the increased absorption of solar energy by the black stripes [42].

Circadian, infradian, and ultradian rhythms influence body temperature and should be considered when planning IRT or interpreting thermographic results [43]. In our study, the IRT measurements were performed in the first part of the day.

Thermography has proven to be valuable in human and equine medicine, as well as in diagnosing musculoskeletal injuries in companion animals [44–46]. Unlike other imaging techniques, it is non-invasive, it does not require anesthesia, and it does not expose the patient to radiation [24,44]. Numerous studies have highlighted thermography's ability to detect changes in thermal patterns prior to the appearance of clinical or radiographic signs [47].

The results obtained indicate an elevated temperature in the thoraco-lumbar region in dogs with intervertebral disc extrusion compared to healthy dogs, with these temperature changes associated with local inflammation and muscular spasms. Further investigations are needed to (1) compare the temperature on the left and right sides of the spine to identify correlations with the lateralization of disc extrusion, (2) compare the temperatures observed in disc extrusion pathologies with those associated with other intervertebral disc diseases, (3) assess differences in local temperatures between dogs with acute disc extrusion and those with chronic disc protrusion, and (4) compare differences in local temperature between haired and non-haired dogs with IVDE.

The limitations of the study included the different breeds of dogs taken into consideration, the different distributions between dog breeds, not cutting the hair, the skin color of the dogs, the absence of myelography to identify a disc extrusion without mineralization, the different distribution of the adipose tissue in the thoraco-lumbar area and the moderate resolution of the thermographic device. The operator was not blinded during thermographic scanning or when marking the selected areas for the study.

5. Conclusions

The local temperature of the thoraco-lumbar area in dogs with IVDE is higher than the local temperature of the same identified area in healthy dogs.

The thermal pattern of the thoraco-lumbar area present differences in dogs with IVDE compared with healthy ones.

We recommend thermographic scanning of the thoraco-lumbar area as a minimally invasive method to distinguish dogs with IVDE from healthy ones, particularly in hairless dogs.

Further investigations are required to differentiate between disc extrusion that exhibits lateralization to the right or left.

Author Contributions: Conceptualization, C.Z. and R.D.; methodology, L.S. and L.C.; software, T.C.; validation, C.Z., V.C. and L.C.; formal analysis, C.R.; investigation, C.Z. and N.P.; resources, C.Z.; data curation, B.S.; writing—original draft preparation, C.Z.; writing—review and editing, L.S. and T.F.; visualization, R.D.; supervision, C.Z.; funding acquisition, R.D. All authors have read and agreed to the published version of the manuscript.

Funding: This study was funded through the project "Increasing the Impact of Excellence Research on Innovation and Technology Transfer Capacity within USAMVB Timisoara", code 6PFE. The project was part of Program 1—Development of the National Research and Development System, Subprogram 1.2—Institutional Performance, under Institutional Development Projects—Excellence Development Projects in RDI.

Institutional Review Board Statement: Ethical review and approval were waived for this study due to its retrospective nature, where data were collected from existing records as clinical cases with the consent of the owners for use in research.

Informed Consent Statement: Informed consent was obtained from all dog owners involved in the study.

Data Availability Statement: All the data obtained for this pilot study are available in the Clinical Register of Companian Animals of the Surgery Clinic.

Acknowledgments: The authors wish to express their gratitude to the animal owners, Dan Bumb, and the students who assisted in the postoperative recovery of the dogs.

Conflicts of Interest: The authors declare no conflicts of interest.

References

1. Coates, J.R. Intervertebral disk disease. *Vet. Clin.* **2000**, *30*, 77–110. [CrossRef] [PubMed]
2. Somerville, M.E.; Anderson, S.M.; Gill, P.J.; Kantrowitz, B.J.; Stowater, J.L. Accuracy of Localization of Cervical Intervertebral Disk Extrusion or Protrusion Using Survey Radiography in Dogs. *J. Am. Anim. Hosp. Assoc.* **2001**, *37*, 563–572. [CrossRef] [PubMed]
3. Fenn, J.; Olby, N.J. Classification of Intervertebral Disc Disease. *Front. Vet. Sci.* **2020**, *7*, 579025. [CrossRef] [PubMed]
4. Wheeler, S.J.; Sharp, N.J. *Small Animal Spinal Diagnosis: Diagnosis and Surgery*, 2nd ed.; Mosby Yearbook: Philadelphia, PA, USA; London, UK, 2005; Volume 33, p. 19.
5. Kranenburg, H.C.; Grinwis, C.M.; Bergknut, N.; Gahrman, N.; Voorhout, G.; Hazewinkel, H.; Meij, B. Intervertebral disc disease in dogs—Part 2: Comparison of clinical, magnetic resonance imaging, and histological findings in 74 surgically treated dogs. *Vet. J.* **2013**, *195*, 164–171. [CrossRef]
6. Jeffery, N.D.; Levine, J.M.; Olby, N.J.; Stein, V.M. Intervertebral disk degeneration in dogs: Consequences, diagnosis, treatment, and future directions. *J. Vet. Intern. Med.* **2013**, *27*, 1318–1333. [CrossRef]
7. Spitzbarth, I.; Sarah, A.M.; Stein, V.M.; Levine, J.M.; Kurl, B.; Gerhauser, I.; Baumgartner, W. Current Insights Into the Pathology of Canine Intervertebral Disc Extrusion-Induced Spinal Cord Injury. *Front. Vet. Sci.* **2020**, *7*, 595796. [CrossRef]
8. Mayhew, P.D.; McLear, R.C.; Ziemer, L.S.; Wiliam, T.N.; Russel, K.N.; Shofer, F.S.; Kapatkin, A.S.; Smith, G.K. Risk factors for recurrence of clinical signs associated with thoracolumbar intervertebral disk herniation in dogs: 229 cases (1994–2000). *J. Am. Vet. Med. Assoc.* **2004**, *225*, 1231–1236. [CrossRef]
9. Prindeze, N.J.; Fathi, P.; Mino, M.J.; Mauskar, N.A.; Travis, T.E.; Paul, D.W.; Moffatt, L.T.; Shupp, J.W. Examination of the early diagnostic applicability of active dynamic thermography for burn wound depth assessment and concept analysis. *J. Burn. Care Res.* **2015**, *36*, 626–635. [CrossRef]
10. Drost, W.T.; Love, N.E.; Berry, C.R. Comparison of radiography, myelography and computed tomography for the evaluation of canine vertebral and spinal cord tumors in sixteen dogs. *Vet. Radiol. Ultrasound* **2017**, *37*, 28–33. [CrossRef]
11. Noyes, J.A.; Thomoscky, S.A.; Chen, A.V.; Owen, T.J.; Fransson, B.A.; Carboneau, K.J.; Matthew, S.M. Magnetic resonance imaging versus computed tomography to plan hemilaminectomies in chondrodystrophic dogs with intervertebral disc extrusion. *Vet. Surg.* **2017**, *46*, 1025–1031. [CrossRef]
12. Love, E.; Berry, L. Tomography for the evaluation of canine vertebral and spinal cord tumors in sixteen dogs. *Vet. Radiol. Ultrasound* **1996**, *37*, 28–33.
13. Emery, L.; Hecht, S.; Sun, X. Investigation of parameters predicting the need for diagnostic imaging beyond computed tomography in the evaluation of dogs with thoracolumbar myelopathy: Retrospective evaluation of 555 dogs. *Vet. Radiol. Ultrasound* **2017**, *59*, 147–154. [CrossRef] [PubMed]
14. Kirberger, R.M.; Roos, C.J.; Lubbe, A.M. The radiological diagnosis of thoracolumbar disc disease in the dachshund. *Vet. Radiol. Ultrasound* **1992**, *33*, 255–261. [CrossRef] [PubMed Central]
15. Purdoiu, R.C.; Ashur, R.; Condor, L.; Lăcătuș, R. Computed Tomography Findings in Spinal Compression in 196 Dogs. *Bull. Univ. Agric. Sci. Vet. Med. Cluj-Napoca. Vet. Med.* **2018**, *75*, 46. [CrossRef]
16. Cooper, J.J.; Young, B.D.; Grifin, J.F.; Fosgate, G.T.; Levine, J.M. Comparison between noncontrast computed tomography and magnetic resonance imaging for detection and characterization of thoracolumbar myelopathy caused by intervertebral disk herniation in dogs. *Vet. Radiol. Ultrasound* **2013**, *55*, 182–189. [CrossRef]
17. Lim, C.; Kweon, O.; Choi, M.; Choi, M.; Yoon, J. Veterinary Science Computed tomographic characteristics of acute thoracolumbar intervertebral disc disease in dogs. *J. Vet. Sci.* **2010**, *11*, 73–79. [CrossRef]
18. Lim, J.; Yoon, Y.; Hwang, T.; Lee, H.C. Novel vertebral computed tomography indices in normal and spinal disorder dogs. *J. Vet. Sci.* **2018**, *19*, 296–300. [CrossRef]
19. Redaelli, V.; Tanzi, B.; Luzi, F.; Stefanello, D.; Proverbio, D. Original Use of thermographic imaging in clinical diagnosis of small animal: Preliminary notes. *Ann. Ist. Super. Sanita* **2014**, *50*, 140–146. [CrossRef]
20. Infernuso, T.; Loughin, C.A.; Marino, D.J.; Umbaugh, S.E.; Solt, P.S. Thermal Imaging of Normal and Cranial Cruciate Ligament-Deficient Stifles in Dogs. *Vet. Surg.* **2010**, *39*, 410–417. [CrossRef]
21. Grossbard, B.P.; Loughin, C.A.; Marino, D.J. Medical Infrared Imaging (Thermography) of Type I Thoracolumbar Disk Disease in Chondrodystrophic Dogs. *Vet. Surg.* **2014**, *43*, 869–876. [CrossRef]
22. Liu, H.; Zhu, Z.; Jin, X.; Huang, P. The diagnostic accuracy of infrared thermography in lumbosacral radicular pain: A prospective study. *J. Orthop. Surg. Res.* **2024**, *19*, 409. [CrossRef] [PubMed]
23. Dudek, K.; Soroko, M.; Wanda, G.; Howell, K.; Zieli, P.; Eberhardt, M. Changes in Body Surface Temperature Associated with High-Speed Treadmill Exercise in Beagle Dogs Measured by Infrared Thermography. *Animals* **2021**, *11*, 2982. [CrossRef] [PubMed]
24. Alves, J.C.; Santos, A.; Jorge, P.; Lavrador, C.; Carreira, L.M. Evaluation of digital thermography imaging to assess and monitor treatment of police working dogs with naturally occurring hip osteoarthritis. *Animals* **2024**, *14*, 696. [CrossRef] [PubMed]
25. Repac, J.; Alvarez, L.X.; Lamb, K.; Gillette, R.L. Evaluation of Thermographic Imaging in Canine Hindlimb Muscles After 6 Min of Walking—A Pilot Study. *Front. Vet. Sci.* **2020**, *7*, 224. [CrossRef]

26. Kwon, C.J.; Brundage, C.M. Quantifying body surface temperature differences in canine coat types using infrared thermography. *J. Therm. Biol.* **2019**, *82*, 18–22. [CrossRef]
27. Lappalainen, A.K.; Vaittinen, E.; Junnila, J.; Laitinen-vapaavuori, E. Intervertebral disc disease in Dachshunds radiographically screened for intervertebral disc calcifications. *Acta Vet. Scand.* **2014**, *56*, 89. [CrossRef]
28. Crawford, A.H.; De Decker, S. Clinical presentation and outcome of dogs treated medically or surgically for thoracolumbar intervertebral disc protrusion. *Vet. Rec.* **2017**, *180*, 569. [CrossRef]
29. Mota-Rojas, D.; Ogi, A.; Villanueva-García, D.; Hernández-Ávalos, I.; Casas-Alvarado, A.; Domínguez-Oliva, A.; Lendez, P.; Ghezzi, M. Thermal Imaging as a Method to Indirectly Assess Peripheral Vascular Integrity and Tissue Viability in Veterinary Medicine: Animal Models and Clinical Applications. *Animals* **2024**, *14*, 142. [CrossRef]
30. Turner, T.A. Diagnostic thermography. *Vet. Clin. N. Am. Equine Pract.* **2001**, *17*, 95–114. [CrossRef]
31. Lahiri, B.B.; Bagavathiappan, S.; Jayakumar, T.; Philip, J. Medical applications of infrared thermography: A review. *Infrared Phys. Technol.* **2012**, *55*, 221–235. [CrossRef]
32. Ogon, I.; Takebayashi, T.; Takashima, H.; Morita, T.; Terashima, Y.; Yoshimoto, M.; Yamashita, T. Imaging diagnosis for intervertebral disc. *JOR Spine* **2020**, *3*, 1066. [CrossRef] [PubMed]
33. Besalti, O.; Pekcan, Z.; Sirin, S.; Erbas, G. Magnetic resonance imaging findings in dogs with thoracolumbar intervertebral disk disease: 69 cases (1997–2005). *J. Vet. Clin.* **2023**, *40*, 38–43. [CrossRef] [PubMed]
34. Müller, M.K.; Ludewig, E.; Oechtering, G.; Scholz, M.; Flegel, T. The vacuum phenomenon in intervertebral disc disease of dogs based on computed tomography images. *J. Small Anim. Pract.* **2013**, *54*, 253–257. [CrossRef] [PubMed]
35. Burke-Smith, A.; Collier, J.; Jones, D.I. A comparison of non-invasive imaging modalities: Infrared thermography, spectrophotometric intracutaneous analysis and laser Doppler imaging for the assessment of adult burns. *Burns* **2015**, *41*, 1695–1707. [CrossRef]
36. Zhang, H.-Y.; Kim, Y.-S.; Cho, Y.-E. Thermatomal changes in Cervical Disc Herniation. *Yonsei Med. J.* **1999**, *40*, 401–412. [CrossRef]
37. McManus, C.; Tanure, C.B.; Peripolli, V.; Seixas, L.; Fischer, V.; Gabbi, A.M.; Menegassi, S.R.O.; Stumpf, M.T.; Kolling, G.J.; Dias, E.; et al. Infrared thermography in animal production: An overview. *Comput. Electron. Agric.* **2016**, *123*, 10–16. [CrossRef]
38. Zaha, C.; Schuszler, L.; Dascalu, R.; Nistor, P.; Florea, T.; Rujescu, C.; Sicoe, B.; Igna, C. Thermographic Image of the Hoof Print in Leisure and Cross-Country Warmblood Horses: A Pilot Study. *Vet. Sci.* **2023**, *10*, 470. [CrossRef]
39. Casas-Alvarado, A.; Martínez-Burnes, J.; Mora-Medina, P.; Hernández-Avalos, I.; Domínguez-Oliva, A.; Lezama-García, K.; Gómez-Prado, J.; Mota-Rojas, D. Thermal and Circulatory Changes in Diverse Body Regions in Dogs and Cats Evaluated by Infrared Thermography. *Animals* **2022**, *12*, 789. [CrossRef]
40. Ishioka, K.; Okumura, M.; Sagawa, M.; Nakadomo, F.; Kimura, K.; Saito, M. Computed tomographic assessment of body fat in beagles. *Vet. Radiol. Ultrasound* **2005**, *46*, 49–53. [CrossRef]
41. Loughin, C.A.; Marino, D.J. Evaluation of thermographic imaging of the limbs of healthy dogs. *Am. J. Vet. Res.* **2007**, *68*, 1064–1069. [CrossRef]
42. Horvath, G.; Pereszlenyi, A.; Szaz, S.; Barta, A. Experimental evidence that stripes do not cool zebras. *Sci. Rep.* **2018**, *8*, 9351. [CrossRef] [PubMed]
43. Steward, M.; Webster, J.R.; Schaefer, A.; Cook, N.J. Infrared thermography as a non-invasive tool to study animal welfare. *Anim. Welf.* **2005**, *14*, 319–325. [CrossRef]
44. Eddy, A.L.; Van Hoogmoed, L.M.; Snyder, J.R. The role of thermography in the management of equine lameness. *Vet. J.* **2001**, *162*, 172–181. [CrossRef] [PubMed]
45. Purohit, R.C.; McCoy, M.D. Thermography in the diagnosis of inflammatory processes in the horse. *Am. J. Vet. Res.* **1980**, *41*, 1167–1174.
46. Vaden, M.F.; Purohit, R.C.; McCoy, M.D.; Vaughan, J.T. Thermography: A technique for subclinical diagnosis of osteoarthritis. *Am. J. Vet. Res.* **1980**, *41*, 1175–1179.
47. Varju, G.; Pieper, C.F.; Renner, J.B.; Kraus, V.B. Assessment of hand osteoarthritis: Correlation between thermographic and radiographic methods. *Rheumatology* **2004**, *43*, 915–919. [CrossRef]

life

MDPI

Case Report

Non-Invasive Paraclinical Diagnosis of Hepatocutaneous Syndrome in a Dog

Anamaria-Hortensia Strichea [1], Ștefania Livia Hreniuc [2] and Gheorghe Solcan [1,*]

[1] Internal Medicine Clinic, Faculty of Veterinary Medicine, Iași University of Life Sciences, 8 M. Sadoveanu Alley, 700489 Iași, Romania; anamaria.strichea@iuls.ro

[2] Diagnostic Imaging Clinic, Faculty of Veterinary Medicine, Iași University of Life Sciences, 8 M. Sadoveanu Alley, 700489 Iași, Romania; hreniucstefania@gmail.com

* Correspondence: gsolcan@uaiasi.ro

Abstract: A 14-year-old, female spayed Bichon Maltese with no other known previous pathologies was presented for dermatological examination after being referred from a private clinic with the suspicion of generalized, treatment-resistant demodicosis. Upon presentation and clinical examination, multiple deep skin scrapings were performed, returning negative parasitological results. Complete blood count and serum biochemistry revealed mild hepatic damage. Abdominal ultrasound revealed an abnormal echostructure of the liver displaying a honeycomb or Swiss cheese-like pattern, reported as pathognomonic for hepatocutaneous syndrome. The owner declined any further paraclinical examination, including skin biopsy and histopathological examination, requesting a treatment protocol that could be pursued at home, considering the age of the dog and its reactive behavior during the examination. The present case report highlights a non-invasive method of diagnosing the hepatocutaneous syndrome in a dog by clinical examination, routine blood testing, and ultrasound assessment of the abdomen, in the absence of the possibility of doing a histopathological diagnosis.

Keywords: hepatocutaneous syndrome; necrolytic migratory erythema; superficial necrolytic dermatitis

1. Introduction

Superficial necrolytic dermatitis is an uncommon and frequently lethal condition in dogs, often linked to pancreatic neuroendocrine neoplasia and hepatocutaneous syndrome [1,2]. In human medicine, necrolytic migratory erythema is a term used to characterize the skin rash observed in individuals with a pancreatic tumor that secretes glucagon (glucagonoma) or occasionally in cases of hepatic cirrhosis and some gastrointestinal disorders. This rash has also been observed in dogs and cats, often in conjunction with liver disease. Other terms such as hepatocutaneous syndrome, superficial necrolytic dermatitis, and metabolic epidermal necrosis are frequently employed to refer to the same condition [3,4].

The precise pathogenesis remains uncertain, but it is believed that elevated gluconeogenesis due to hyperglucagonemia (associated with pancreatic tumors) or heightened hepatic breakdown of amino acids (in cases of chronic liver disease) leads to decreased plasma amino acid levels and depletion of epidermal proteins. This depletion is what causes the skin lesions characteristic of superficial necrolytic dermatitis. While uncommon in dogs and rare in cats, the condition is most frequently observed in older animals [5].

The lesions initially manifest in regions of the body with high cell turnover, such as mucocutaneous junctions, the face, the footpads, and areas prone to pressure. These lesions typically consist of crusts and ulcers accompanied by reddened skin surrounding the affected area. Additionally, most affected dogs exhibit non-regenerative anemia, mild elevation in blood sugar levels, heightened serum liver enzyme levels, and a liver appearance resembling a "honeycomb" pattern under abdominal ultrasonography [3,6–8].

Although there are some reports regarding concurrent pancreatic changes [9], they were not observed in the present case.

2. Case Description

A 14-year-old Bichon Maltese was presented for dermatological examination to the Internal Medicine Clinic of the Faculty of Veterinary Medicine of Iași. The dog had previously undergone examination and a treatment protocol for demodicosis at a private clinic. The treatment included three consecutive monthly external antiparasitic treatments: the first two with an oral formulation (Nexgard Spectra, Boehringer Ingelheim International GmbH, Ingelheim am Rhein, Germany) containing afoxolaner (37.50 mg)/milbemycin oxime (7.50 mg) as active substances, and the third with a spot-on application of an imidacloprid (100 mg)/permethrin (500 mg) external antiparasitic (Advantix, Elanco, IN, USA). Additionally, the dog had been bathed once a week for the previous four weeks using a sebolytic shampoo (Sebolytic, Virbac, Suite, TX, USA). The animal had not received any drugs potentially responsible for drug-induced skin lesions [3,5,6]. Despite the antiparasitic treatments and topical therapy, no positive results were observed, prompting a referral for a second opinion.

The owners reported that the dermatological condition had begun approximately three months prior to the examination, and although some lesions appeared to have healed, new lesions emerged in different areas. When questioned about pruritus, the owners rated it as 5/10 on the PVAS scale [10]. According to the owners, the dog had a good appetite and no other known health issues apart from the cutaneous lesions, which were observed to be painful as evidenced by increased sedentarism and evident mobility challenges attributable to inflammation affecting the paws. The dog's diet consisted of a mixture of human food and commercial wet dog food.

During physical examination, the patient appeared anxious and displayed reactive behavior towards the medical team which led to handling exclusively by the owners. The body temperature was 39.3 °C, with pale pink oral and conjunctival mucosa, and a CRT of less than 2 s. No abdominal distress was noted upon palpation. The skin and coat emitted a seborrheic odor, and multiple lesions were observed on the skin surface, characterized as follows:

- exfoliative, crusted, and ulcerated cutaneous lesions with purulent secretion localized periocular (Figures 1a and 2a,b), on the nasal planum (Figure 1a,b), symmetrically in the inguinal region (Figure 3a,b and Figure 4a) and on bilateral pinnae (Figure 4b);
- severe four-limb pododermatitis characterized by exfoliation of the plantar pads accompanied by adjacent purulent secretion (Figure 5a) and interdigital pustules with associated secretion (Figure 5b).

Multiple deep skin scrapings were performed, yielding negative results, thereby excluding parasite infestation. Skin and ear cytology revealed the abundant presence of neutrophils (1–3 neutrophils/100× objective field for both ear swabs and 5–10 neutrophils/100× objective field for the skin cytology) and cocci (5–10 extracellular cocci/100× objective field for both ear swabs, 1–5 intracellular cocci/100× objective field for both ear swabs, >10 extracellular cocci/100× objective field of skin cytology, and 5–10 intracellular cocci/100× objective field of skin cytology).

Blood samples were collected for a complete blood count and serum biochemistry analysis. The complete blood count performed using Abaxis VetScan HM5 (Zoetis Services, Parsippany, NJ, USA) showed platelet counts exceeding the upper limit of the reference range, while lymphocyte (LYM), mean corpuscular hemoglobin (MCH), and mean corpuscular volume (MCV) values were at the lower end of the reference range. All of the other parameters were in normal range (Figure S1).

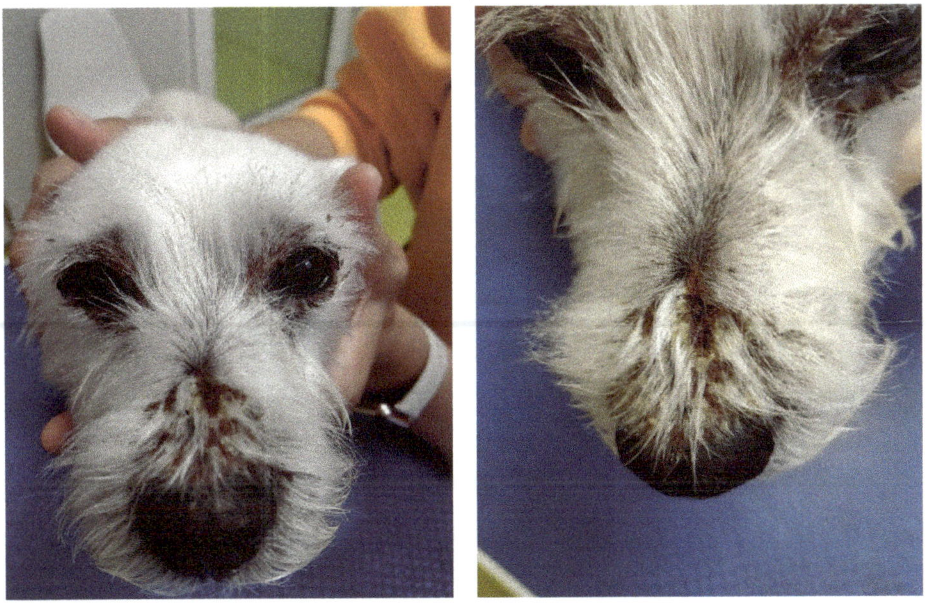

(**a**) (**b**)

Figure 1. (**a**) Crusted lesions localized periocular and on the nasal planum. (**b**) Crusts situated on the nasal planum.

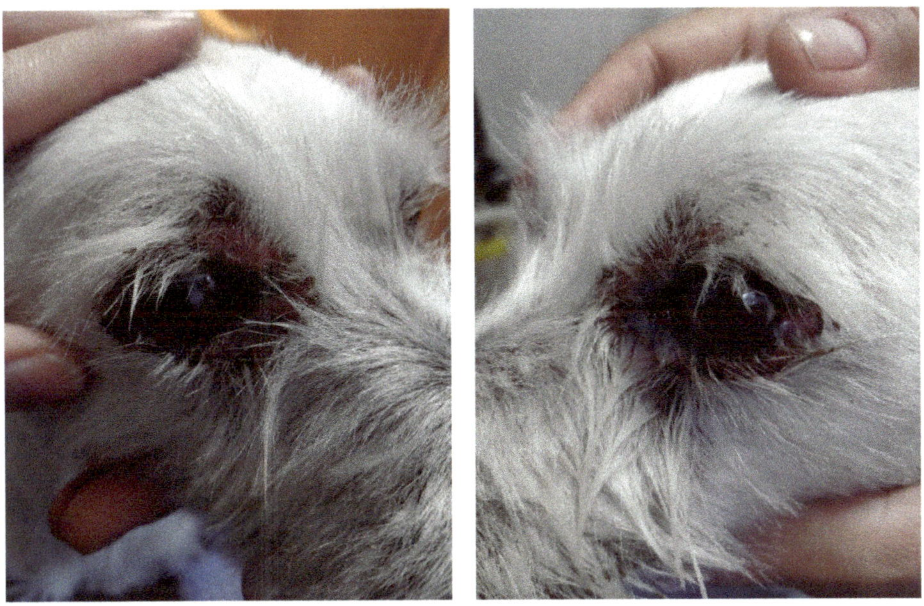

(**a**) (**b**)

Figure 2. (**a**) Erosions in the periocular area of the right eye accompanied by adjacent alopecia. (**b**) Erosions in the periocular region of the left eye accompanied by adjacent alopecia and purulent discharge.

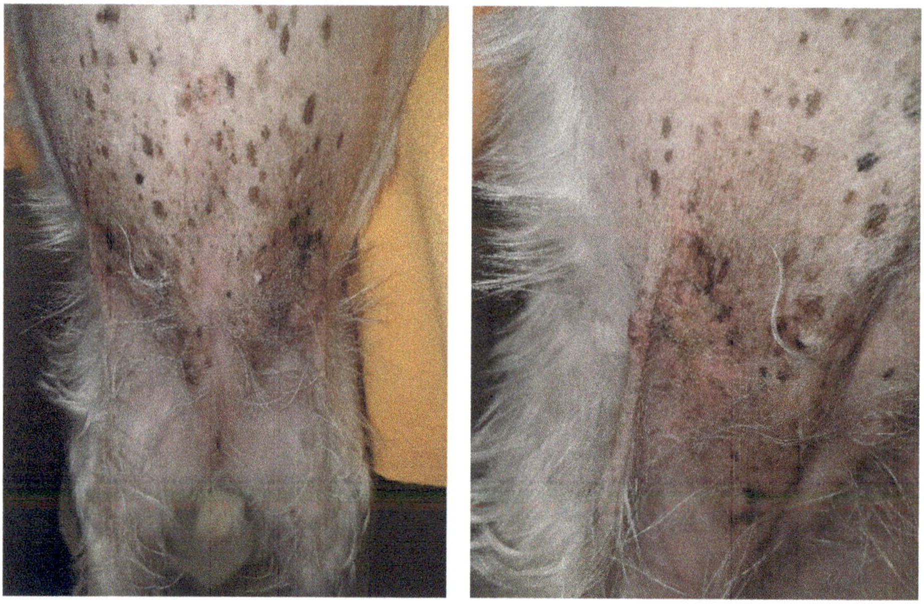

(a) (b)

Figure 3. (**a**) Symmetrical appearance of the lesions situated in the inguinal region. (**b**) Ulcerated lesion covered with crust and adjacent erythema observed in the right inguinal area.

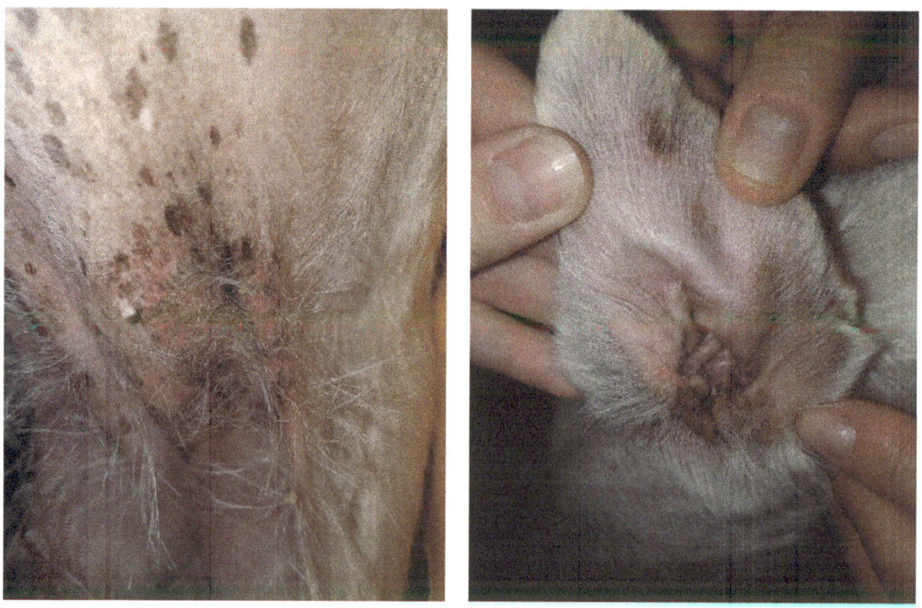

(a) (b)

Figure 4. (**a**) Ulcerated lesion with erythema observed in the left inguinal area. (**b**) Round erythematous and crusty lesion located near the margin of the left pinna, along with the presence of erythema and crusts at the entrance of the ear canal.

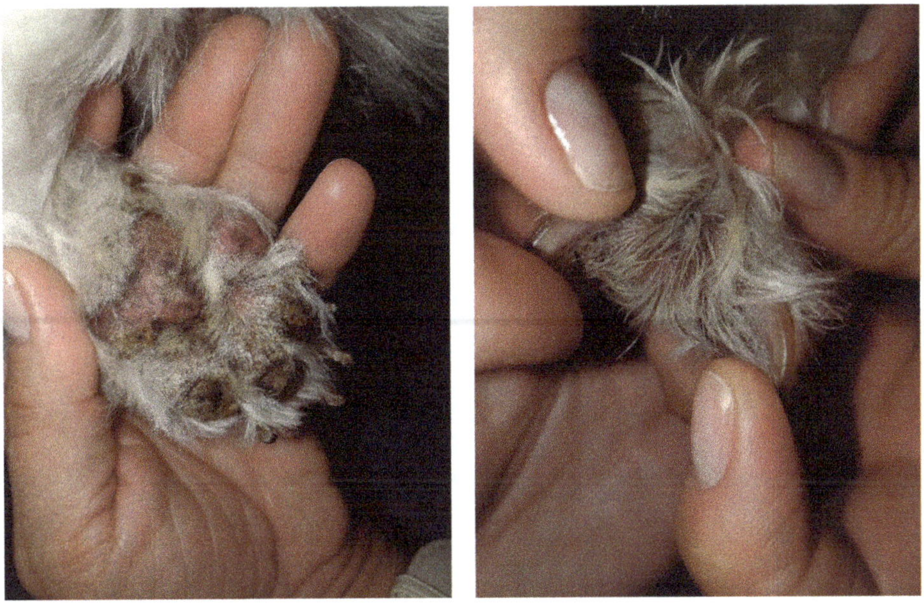

(a) (b)

Figure 5. (**a**) Severe pododermatitis with the presence of purulent discharge and crusts observed between the digital pads. (**b**) Interdigital presentation characterized by purulent discharge, with the claws appearing normal.

Serum biochemistry analysis performed using Abaxis VetScan VS2 (Zoetis Services, Parsippany, NJ, USA)—VetScan Comprehensive Diagnostic Profile indicated mild hepatic damage, with elevated levels of alkaline phosphatase (ALP) at 281 U/L and alanine aminotransferase (ALT) at 133 U/L, as well as glucose (GLU) at 115 mg/dL and potassium (K+) at 5.9 mmol/L, with all other parameters falling within normal range limits (Figure S2).

Given the exclusion of parasitic infestation, the presentation pattern of the skin lesions, and the findings from the complete blood count and serum biochemistry analysis, an abdominal ultrasound was performed, using the General Electric LOGIQ V5 Expert Ultrasound Machine (GE Medical Systems, Wuxi, Jiangsu, China) equipped with a 7.5–10 MHz microconvex transducer and a 7–13 MHz linear transducer. The abdomen was examined in a counter clockwise direction starting from the urinary bladder.

The urinary bladder appeared mildly distended with anechogenic content and normal wall layering, though thickened (0.36 cm), and presenting a slightly irregular luminal lining, with ultrasonographic findings being suggestive for cystitis.

Both kidneys presented with normal echostructure, echogenicity and size, smooth outline, and no visible cystic/nodular structures or calculi, but there were multiple calyceal and cortical punctiform hyperattenuating foci observed, which were believed to be mineralization in absence of a histopathological diagnosis (Figure S3).

The stomach had a normal wall layering and measurement with reduced content and a hyperreflective structure with a posterior acoustic shadowing-compatible with bone ingestion which was confirmed by the owner (Figure S4). The small intestine had normal wall layering and measurements and normal peristaltic movement and the large intestine presented with normal wall layering but slightly thickened (0.32 cm). The pancreas appeared normal, with no adjacent mesenteric reaction or free fluid.

The spleen appeared with predominantly normal echostructure and echogenicity, normal size (1.17 cm short axis at the level of the splenic hilum). Cranial to the splenic hilum there was a poorly delimited area, approximately 0.75 × 0.42 cm in size, non-homogeneous, showing both hyperechoic and anechoic regions, with an absence of Doppler signal, discretely deforming the splenic contour. Cranial to these there were found two other poorly defined areas, hyperechoic compared to the rest of the parenchyma, the largest measuring 0.33 × 0.25 cm. The ultrasonographic differential diagnosis established was lipomas/myelolipomas/neoplasia and focal areas of dystrophic mineralization/neoplasia (Figure S5).

The liver appeared of normal size with non-homogeneous parenchyma, with multiple hyperechogenic strands diffusely dispersed throughout the parenchyma giving it a "Swiss cheese" appearance, with a slightly irregular outline and poorly distinguishable vascular walls (Figures 6 and 7a,b). The gall bladder had a fine hyperechoic wall with anechoic content accompanied by hypoechoic, mobile, gravitationally dependent sediment in reduced quantity.

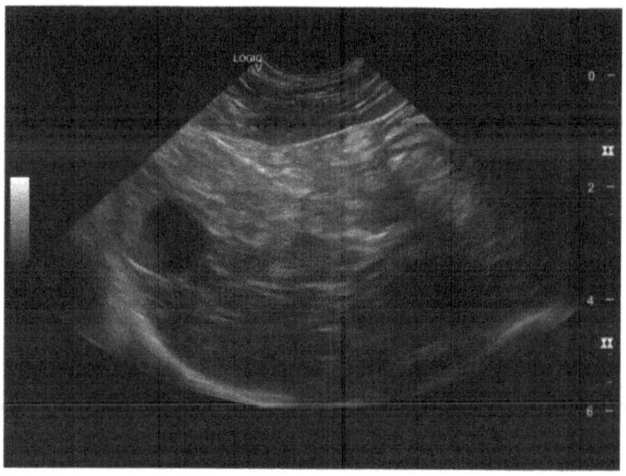

Figure 6. Liver and gallbladder—multiple hyperechogenic strands diffusely dispersed throughout the parenchyma—"Swiss cheese" appearance and gallbladder sediment (microconvex transducer).

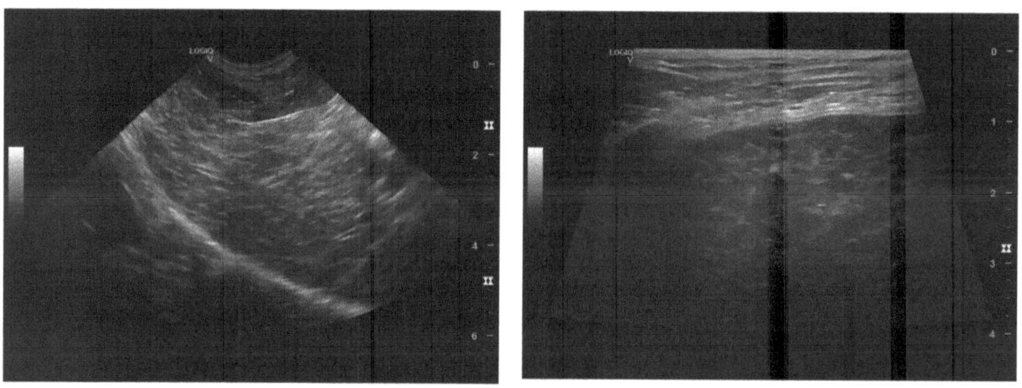

(**a**) (**b**)

Figure 7. (**a**) Liver—non-homogeneous parenchyma (microconvex transducer). (**b**) Liver multiple hyperechogenic strands diffusely dispersed throughout the parenchyma—"Swiss cheese" appearance (linear transducer).

The histopathological examination of the skin and liver was proposed to the owners, but they declined the fine needle aspiration biopsy of the liver and the punch biopsy of the skin, considering the invasiveness of both sampling techniques and the need for anesthesia.

The owners also declined any other parenteral treatment, requesting a protocol that could be administered by them at home. The recommended treatment consisted of a liquid oral supplement of amino acids (Rx-amino B-plex, RX Vitamins Inc., Suite, TX, USA—2 mL twice a day for 30 days), an essential fatty acid, vitamin, and zinc supplement (VetoSkin 300 mg twist off, Vetexpert, Łomianki, Poland—1 capsule/day for 30 days) and a hepatic support supplement (FOR Liver, Crida Pharm, Bucharest, Romania—1/2 pill/day for 30 days). The recommended diet consisted of a homecooked protein source (egg whites—one/day) and a commercial hepatic diet (Hill's I/D Liver Care, Hill's, Topeka, KS, USA). A 4% chlorhexidine shampoo (Clorhexyderm ICF, PetMart, Bucharest, Romania—2 baths/week for 4 weeks) and an ointment containing a combination of antibiotics (tetracycline, erythromycin, neomycin) and prednisolone (Mibazon, Antibiotice, Iași, Romania) one application daily for 28 days) were prescribed for the topical treatment of the pyoderma.

3. Discussion

This case report suggests that in dogs with characteristic skin lesions, a comprehensive diagnostic approach including abdominal ultrasound assessment of the liver and pancreas, as well as complete blood count and serum biochemistry, allows for a noninvasive diagnosis of hepatocutaneous syndrome, particularly pertinent in cases where the owners decline further invasive paraclinical examinations.

The dermatological differential diagnosis of hepatocutaneous syndrome includes autoimmune skin diseases (pemphigus foliaceus or vulgaris, systemic lupus erythematosus), drug eruption, and zinc-responsive dermatosis [3,5] as the pattern of the skin lesions could be similar. The drug cutaneous eruption diagnosis was excluded since the anamnesis did not highlight any intake of drugs that could trigger a reaction. Since zinc-responsive dermatitis develops mostly in puppies [11] as an improper feeding consequence, it was also excluded from the differential list.

Pemphigus complex is a group of autoimmune skin disease that are characterized by blister and pustule formation due to the loss of adhesion between the keratinocytes [3,5,12–14]. In pemphigus foliaceus, the lesions are pustular at the beginning and by pustule breaking, crusts and pyoderma are formed. The location of skin lesions in dogs are reported as follows: trunk, inner pinnae, dorsal muzzle, foot pads, periocular area, outer pinnae, planum nasale, interdigital area, lips, perianal area, and mucous membranes [14]. The pattern of the lesions is described as bilateral and symmetrical [13]. Performing direct cytology from the pustules can offer a quick diagnosis of pemphigus foliaceus if acantholytic cells are found [14]. The dog from our case report displayed crusted lesions in which acantholytic cells were not present, but a final diagnosis could not be established by dermatological examination and direct cytology alone. Pemphigus vulgaris is an autoimmune blistering dermatological pathology that especially affects the mucosae and mucocutaneous junction and can affect also the skin. The lesions evolve from blisters (vesicle or bullae) to erosions that generate pain, especially when the development is in the oral cavity [15]. In a study that corroborated the data from 54 cases of canine pemphigus vulgaris, it was concluded that the majority of cases displayed a mucosal or mucocutaneous phenotype in which the lesions were located on the lips/oral cavity. Only a few cases displayed skin lesions alone (2 out of 54) [15].

The clinical presentation of cutaneous lupus erythematosus is highly variable, ranging from minor, mild alopecic scarring lesions to extensive ulceration. Some patients exhibit pruritic seborrheic dermatitis, while others present with mucocutaneous ulceration, which can disseminate to other body regions. Depigmentation of the nose or periocular area may occur, alongside erythema, ulceration, and crusting of the nose, which is characteristic of discoid lupus erythematosus. Focal ulceration of the footpads, suggestive of vasculitis, may also be present, along with more diffuse crusting pad lesions [3].

Since the pathogenesis of the pemphigus complex diseases is related only to the immune mediated conflict at the skin/mucosal level, even if the location of the skin lesions are similar to those from pemphigus foliaceus and lupus erythematosus, in the present case report, the involvement of the liver lesions is definitory in orientating the diagnosis towards hepatocutaneous syndrome.

Previously reported abnormal clinical pathology findings in dogs diagnosed with superficial necrolytic dermatitis or hepatocutaneous syndrome include anemia, microcytosis, elevated alkaline phosphatase activity, and hypoalbuminemia. Elevated alkaline phosphatase activity was consistently noted in cases of aminoaciduric canine hypoaminoacidemic hepatopathy syndrome (ACHES), irrespective of the presence of cutaneous lesions. Anemia and microcytosis were particularly pronounced in dogs with fulminant ACHES, indicating a potential correlation between superficial necrolytic dermatitis and these hematological alterations. Profiling of plasma and urine amino acids offers a noninvasive approach for disease confirmation. As previously demonstrated, urine amino acid profiling normalized to creatinine, revealing lysinuria, presents a novel diagnostic parameter that could enhance the diagnostic efficacy of plasma amino-acid profiling [16].

The patient in this study exhibited subtle modifications in the complete blood count, specifically lymphocytes (LYM), mean corpuscular volume (MCV), and mean corpuscular hemoglobin (MCH) at the lower limit of the reference range, while platelets (PLT) exceeded the upper limit of the reference range, thus suggesting the presence of microcytic anemia. Serum biochemistry indicated values above the limit for alkaline phosphatase (ALP) and alanine aminotransferase (ALT), with albumin levels within physiological limits. A limitation of this case report is the inability to quantitatively determine plasma and urinary amino acid values due to external reasons (infrastructure liabilities of the laboratory and the lack of other options in the geographical area).

The ultrasonographical aspect of the liver as well as the normal size of the organ combined with the Swiss cheese-like appearance of the parenchyma corroborates the diagnosis of hepatocutaneous syndrome. Another differential diagnosis in absence of the dermatological findings could have been that of chronic hepatitis/liver cirrhosis, although the liver is reported to be reduced in size in these common conditions.

Historically, superior outcomes for dogs with hepatocutaneous syndrome have been attributed to the administration of intravenous amino acid solutions [17]. Alongside intravenous amino acid infusions, high-protein diets, typically commercial diets supplemented with whey protein, have been a cornerstone of treatment for dogs with hepatocutaneous syndrome [18]. More recently, the combined administration of intravenous lipid with intravenous amino acid infusions was reported to manage hepatocutaneous syndrome in a single dog for 24 months. These observations suggest improved remission and survival in dogs with aminoaciduric canine hypoaminoacidemic hepatopathy syndrome when fed high-protein home-cooked diets [2]. In a separate study, the utilization of a combined treatment approach involving amino acid supplementation and stem cell therapy demonstrated efficacy in a patient diagnosed with hepatocutaneous syndrome. Notably, the dog exhibited survival for a period of 32 months following diagnosis [19].

In the case report presented, the owner of the dog requested a treatment protocol that could be administered at home. The treatment protocol included oral supplements of amino acids, minerals and vitamins, essential fatty acids, and a hepatic support supplement, as well as a commercial hepatic diet supplemented with a home-cooked protein source. Additionally, topical treatment of the pyoderma was prescribed, consisting of using twice a week a 4% chlorhexidine shampoo and an ointment for daily use, containing a combination of antibiotics (tetracycline, erythromycin, neomycin) and an anti-inflammatory (prednisolone). However, the follow-up of the patient could not be conducted as the owner did not attend the 30-day recheck appointment.

The majority of reported outcomes for dogs with hepatocutaneous syndrome are poor, with an average survival of 3–6 months from the time of diagnosis [4,20,21], although sporadic reports indicate survival times exceeding 2 years [2].

It has been hypothesized that dermal lesions manifest as an advanced phase of hepatocutaneous syndrome, and their absence, consequently, does not exclude the presence of the condition [16]. Hence, particularly when dermatologic patients exhibit cutaneous symptoms suggestive of hepatocutaneous syndrome, it is assumed that abdominal ultrasound evaluation serves as an essential non-invasive diagnostic tool and therefore, it was considered that in this situation the histopathological examination of the liver would not be mandatory, once ultrasonographic aspects of the liver are considered pathognomonic [9,22].

4. Conclusions

The hepatocutaneous syndrome of the dog appears to be a complex pathology that can be easily confused with other skin diseases such as pemphigus complex skin diseases, lupus erythematosus, zinc responsive dermatitis, or drug-induced cutaneous reaction and for whose diagnosis the paraclinical examinations are of particular importance. A non-invasive approach of diagnosing the hepatocutaneous syndrome in a dog by clinical examination, routine blood testing, and ultrasound assessment of the abdomen was pursued in the absence of the possibility of performing a histopathological diagnosis.

Ultrasound examination of the abdomen is not a frequently used paraclinical examination technique for the dermatological patients, and it should be considered before other options. We conclude that in patients with clinical symptoms compatible with the previously mentioned pathologies, the first step in establishing a diagnosis should be the exclusion of hepatocutaneous syndrome by non-invasive methods, and unless liver damage is excluded, histopathological examination of the skin should be considered necessary.

Supplementary Materials: The following supporting information can be downloaded at: https://www.mdpi.com/article/10.3390/life14070853/s1. Figure S1: Complete blood count of the patientt; Figure S2: Biochemistry serum analyses of the patient; Figure S3: Spleen- hyperechoic marginal areas (linear transducer); Figure S4: Stomach- hyperreflective structure with a posterior acoustic shadowing (microconvex transducer); Figure S5: Kidney- hyperechoic cortical foci (linear transducer).

Author Contributions: Conceptualization, A.-H.S. and Ș.L.H.; writing—original draft preparation, A.-H.S. and Ș.L.H.; writing—review and editing, A.-H.S. and G.S.; supervision, G.S. All authors have read and agreed to the published version of the manuscript.

Funding: This research received no external funding.

Institutional Review Board Statement: The study was conducted in accordance with the Declaration of Helsinki, and approved by the Ethics Comission of the Faculty of Veterinary Medicine in Iasi within the "Ion Ionescu de la Brad" University of Life Sciences (No. 1749, 20 October 2023).

Informed Consent Statement: Not applicable.

Data Availability Statement: The data present in this study are available within the article.

Acknowledgments: The authors thank Anca Maria Carter Timofte, University of Liverpool, for reviewing and correcting the English version of the manuscript.

Conflicts of Interest: The authors declare no conflicts of interest.

References

1. DeMarle, K.B.; Webster, C.R.L.; Penninck, D.; Ferrer, L. Approach to the Diagnosis of Hepatocutaneous Syndrome in Dogs: A Retrospective Study and Literature Review. *J. Am. Anim. Hosp. Assoc.* **2021**, *57*, 15–25. [CrossRef] [PubMed]
2. Loftus, J.P.; Miller, A.J.; Center, S.A.; Peters-Kennedy, J.; Astor, M. Treatment and Outcomes of Dogs with Hepatocutaneous Syndrome or Hepatocutaneous-associated Hepatopathy. *J. Vet. Intern. Med.* **2022**, *36*, 106–115. [CrossRef] [PubMed]
3. Miller, W.H.; Griffin, C.E.; Campbell, K.L.; Muller, G.H.; Scott, D.W. *Muller & Kirk's Small Animal Dermatology*, 7th ed.; Elsevier: St. Louis, MO, USA, 2013; ISBN 978-1-4160-0028-0.
4. Almendros, A.; Sandy, J.R.; Kirberger, R.M. Hepatocutaneous Syndrome in a Maltese, Diagnosis, Treatment and the Value of CT in the Diagnosis. *Vet. Rec. Case Rep.* **2019**, *7*, e000918. [CrossRef]
5. Hnilica, K.A.; Medleau, L. *Small Animal Dermatology: A Color Atlas and Therapeutic Guide*, 3rd ed.; Elsevier: St. Louis, MO, USA; Saunders: Philadelphia, PA, USA, 2011; ISBN 978-1-4160-5663-8.
6. Ettinger, S.J.; Feldman, E.C.; Côté, E. (Eds.) *Textbook of Veterinary Internal Medicine: Diseases of the Dog and the Cat*, 8th ed.; Elsevier: St. Louis, MO, USA, 2017; ISBN 978-0-323-31211-0.
7. Mattoon, J.S.; Nyland, T.G. *Small Animal Diagnostic Ultrasound*, 3rd ed.; Elsevier: St. Louis, MO, USA, 2015; ISBN 978-1-4160-4867-1.

8. Rhodes, K.H.; Werner, A.H.; Rhodes, K.H. *Blackwell's Five-Minute Veterinary Consult Clinical Companion: Small Animal Dermatology*, 2nd ed.; Blackwell's Five Minutes Veterinary Consult; Wiley-Blackwell: Ames, IA, USA, 2011; ISBN 978-0-8138-1596-1.
9. Jacobson, L.S.; Kirberger, R.M.; Nesbit, J.W. Hepatic Ultrasonography and Pathological Findings in Dogs with Hepatocutaneous Syndrome: New Concepts. *J. Vet. Intern. Med.* **1995**, *9*, 399–404. [CrossRef] [PubMed]
10. Hill, P.B.; Lau, P.; Rybnicek, J. Development of an Owner-Assessed Scale to Measure the Severity of Pruritus in Dogs. *Vet. Dermatol.* **2007**, *18*, 301–308. [CrossRef] [PubMed]
11. Pereira, A.M.; Maia, M.R.G.; Fonseca, A.J.M.; Cabrita, A.R.J. Zinc in Dog Nutrition, Health and Disease: A Review. *Animals* **2021**, *11*, 978. [CrossRef] [PubMed]
12. Lemmens, P.; De Bruin, A.; De Meulemeester, J.; Wyder, M.; Suter, M.M. Paraneoplastic Pemphigus in a Dog. *Vet. Dermatol.* **1998**, *9*, 127–134. [CrossRef] [PubMed]
13. Olivry, T.A. Review of Autoimmune Skin Diseases in Domestic Animals: I—Superficial Pemphigus. *Vet. Dermatol.* **2006**, *17*, 291–305. [CrossRef] [PubMed]
14. Mueller, R.S.; Krebs, I.; Power, H.T.; Fieseler, K.V. Pemphigus Foliaceus in 91 Dogs. *J. Am. Anim. Hosp. Assoc.* **2006**, *42*, 189–196. [CrossRef] [PubMed]
15. Tham, H.L.; Linder, K.E.; Olivry, T. Deep Pemphigus (Pemphigus Vulgaris, Pemphigus Vegetans and Paraneoplastic Pemphigus) in Dogs, Cats and Horses: A Comprehensive Review. *BMC Vet. Res.* **2020**, *16*, 457. [CrossRef] [PubMed]
16. Loftus, J.P.; Center, S.A.; Astor, M.; Miller, A.J.; Peters-Kennedy, J. Clinical Features and Amino Acid Profiles of Dogs with Hepatocutaneous Syndrome or Hepatocutaneous-Associated Hepatopathy. *J. Vet. Intern. Med.* **2022**, *36*, 97–105. [CrossRef] [PubMed]
17. Bach, J.F.; Glasser, S.A. A Case of Necrolytic Migratory Erythema Managed for 24 Months with Intravenous Amino Acid and Lipid Infusions. *Can. Vet. J.* **2013**, *54*, 873–875. [PubMed]
18. Jaffey, J.A.; Backus, R.C.; Sprinkle, M.; Ruggiero, C.; Ferguson, S.H.; Shumway, K. Successful Long-Term Management of Canine Superficial Necrolytic Dermatitis with Amino Acid Infusions and Nutritionally Balanced Home-Made Diet Modification. *Front. Vet. Sci.* **2020**, *7*, 28. [CrossRef] [PubMed]
19. Nam, A.; Han, S.-M.; Go, D.-M.; Kim, D.-Y.; Seo, K.-W.; Youn, H.Y. Long-Term Management with Adipose Tissue-Derived Mesenchymal Stem Cells and Conventional Treatment in a Dog with Hepatocutaneous Syndrome. *J. Vet. Int. Med.* **2017**, *31*, 1514–1519. [CrossRef] [PubMed]
20. Outerbridge, C.A.; Marks, S.L.; Rogers, Q.R. Plasma Amino Acid Concentrations in 36 Dogs with Histologically Confirmed Superficial Necrolytic Dermatitis. *Vet. Dermatol.* **2002**, *13*, 177–186. [CrossRef] [PubMed]
21. March, P.A.; Hillier, A.; Weisbrode, S.E.; Mattoon, J.S.; Johnson, S.E.; DiBartola, S.P.; Brofman, P.J. Superficial Necrolytic Dermatitis in 11 Dogs with a History of Phenobarbital Administration (1995–2002). *J. Vet. Int. Med.* **2004**, *18*, 65–74. [CrossRef]
22. Nyland, T.G.; Barthez, P.Y.; Ortega, T.M.; Davis, C.R. Hepatic Ultrasonographic and Pathologic Findings in Dogs with Canine Surperficial Necrolytic Dermatitis. *Vet. Radiol. Ultrasound* **1996**, *37*, 200–205. [CrossRef]

life

MDPI

Case Report

Electroencephalographic Features of Presumed Hepatic Encephalopathy in a Pediatric Dog with a Portosystemic Shunt—A Case Report

Raluca Adriana Ștefănescu [1], Vasile Boghian [2,*], Gheorghe Solcan [2,*], Mario Darius Codreanu [3] and Mihai Musteata [1]

1 Neurology Service, Faculty of Veterinary Medicine, "Ion Ionescu de la Brad" Iași University of Life Sciences, 700489 Iași, Romania; raluca.stef@yahoo.ro (R.A.Ș.); mihai.musteata@iuls.ro (M.M.)
2 Internal Medicine Clinic, Faculty of Veterinary Medicine, "Ion Ionescu de la Brad" Iași University of Life Sciences, 8 M. Sadoveanu Alley, 700489 Iași, Romania
3 Internal Medicine Clinic, Faculty of Veterinary Medicine, University of Agronomical Sciences and Veterinary Medicine, 105 Independency Spl., 050097 Bucharest, Romania; codveterinary@yahoo.com
* Correspondence: vasile.boghian@iuls.ro (V.B.); gsolcan@uaiasi.ro (G.S.)

Abstract: Hepatic encephalopathy (HE) in dogs is a metabolic disorder of the central nervous system that occurs secondarily to liver dysfunctions, whether due to acquired or congenital causes. A portosystemic shunt is the presence of abnormal communications between the hepatic vessels (portal and suprahepatic veins). As a result of this, the blood brought from the digestive tract through the portal vein bypasses the liver, and the unmetabolized components of the portal bloodstream enter directly into systemic circulation, causing clinical symptoms of metabolic encephalopathy (HE). A 3-month-old Bichon canine patient with a history of seizures secondarily to a portosystemic shunt (PS), confirmed through color Doppler ultrasound exam and computed tomography, was presented for evaluation. The typical electroencephalographic (EEG) traces recorded were characterized by the presence of bilateral symmetrical triphasic waves, resembling non-convulsive status epilepticus. The presence of this EEG pattern is useful in choosing the best therapeutic option in order to not accentuate the HE sings and, consequently, to decrease the mortality risk due to a prolonged status epilepticus.

Keywords: dog; portosystemic shunt; seizures; computed tomography; electroencephalography

1. Introduction

Hepatic encephalopathy (HE) in dogs is a metabolic disorder of the central nervous system that occurs secondarily to liver dysfunction, regardless of whether they have acquired or congenital causes [1–3]. One of the most common causes of HE is the presence of abnormal communications between hepatic vessels (portal vein and suprahepatic vein) that form a portosystemic shunt. As a result of this, the blood brought from the digestive tract through the portal vein bypasses the liver, and the unmetabolized components of the portal bloodstream enter directly into the systemic circulation, causing clinical symptoms of metabolic encephalopathy (ME). Toxic substances, such as ammonia, aromatic amino acids (including phenylalanine, tyrosine, and tryptophan), short-chain fatty acids, mercaptans, and various biogenic amines, indoles, and skatoles have been implicated in causing HE. These compounds have convulsive effects in animals, and can modulate cortical excitability by acting as GABA-A receptor antagonists and NMDA receptor agonists, thus leading to an increase in cortical excitability [4].

Metabolic encephalopathy due to portosystemic shunting causes functional disturbances of CNS activity due to endogenous intoxication.

Secondary hepatic encephalopathy (HE) due to portosystemic shunting is easily suspected if clinical signs appear following food ingestion. Historically, it has been accepted that neurological symptoms manifest 30–90 min after ingestion, and disappear within a few hours, with seizures being the most expressive clinical signs [3]. Suggestive clinical signs, breed predisposition, and additional tests (laboratory tests, along with routine imaging such as ultrasound, or advanced techniques such as computed tomography, etc.) are sufficient to establish the diagnosis of vascular anomaly [4,5]. In reality, neurological clinical signs are present continuously, but if their intensity is very low, the clinical signs may be subtle and expressed only through mild, nonspecific cortical inhibition or changes in behavior [4]. Especially in those animals, assessing the neurological impact of this vascular anomaly is challenging.

Electroencephalographic (EEG) examination is a procedure that allows the real-time assessment of the bioelectric activity of the cerebral cortex [6–8] and the occurrence of superimposed generalized spikes or sharp waves on a slow wave background is a common finding [9]. In humans, understanding the specific EEG patterns in patients with portosystemic shunting is helpful for clinicians in establishing a diagnosis, assessing the severity of cerebral homeostasis disruption, and monitoring the therapeutic effects of medication administered to patients up to the time of surgery (if feasible) [10]. In early stages of HE, EEG traces are characterized by an increase only of beta activity [11] or alpha activity [12] into the frontal lobes, evolving in more severe stages of HE, to bioelectric transients as triphasic waves and bursts of intermittent rhythmic delta activity are seen both in dogs and humans [8,10].

Rapid identification of the etiology in a seizuring or unconscious dog and specific treatment initiation should be rapidly performed, as long as a prolonged unconscious status is associated with a poor outcome [13,14]. This goal may be difficult to achieve, especially in patients with structural brain events who needs advance imaging equipment to be diagnosed, or in patients with nonconvulsive status epilepticus due to either a systemic metabolic disease or a primary structural brain lesion. In such scenarios, identification of a specific pattern of the bioelectric cerebral activity together with routine lab results might offer the clinician a quick indication over both the etiology of the encephalopathies and also the patient's prognosis and potential treatment efficacy.

The aim of the present study is to present the typical EEG recorded in a pediatric canine patient with a presumed HE secondarily to an extrahepatic PS.

2. Case Description

An intact, crossbred male Bichon Frisé, 3 months old and weighing 2 kg, was admitted to the Neurology Service of the Veterinary Teaching Hospital of FMV, Iași, for evaluation of epileptiform seizures with a 4-day history. Seizure activity was characterized by loss of consciousness, rolling and pedaling movements, hypersalivation, and spontaneous emission of urine, which occurred mainly postprandial. Upon admission, the patient was presented with an inconstant opisthotonus with thoracic limb extension and pelvic limb flexion beneath the body (mimicking a decerebellate rigidity), along with severe cranial nerve deficits: normal blink reflex, but delayed and incomplete photomotor reflex, delayed physiological nystagmus, bilateral ventral positional strabismus, bilateral absent menace response, and an absent response to bilateral nasal cantus stimulation. A diffuse intracranial multifocal condition (forebrain, brainstem, and potentially cerebellum) was suspected, and a metabolic, toxic, inflammatory etiology was privileged.

Laboratory tests showed a hemogram with neutrophilic leukocytosis, and serum biochemistry revealed abnormalities in routine hepatic markers, including AST 130 U/L (normal range of 14–45 U/L), ALP 207 (normal range of 20–150 U/L), and GGT 9 (normal range of 0–7 U/L). An abdominal ultrasound examination (My Lab 40 ultrasound machine, Esaote Pie Medical, Maastricht, The Netherlands) revealed a congenital portosystemic shunt (Figure 1) (in comparison to a physiological aspect, Figure S1), for which the patient was recommended to undergo bile acid testing and advanced imaging (abdominal CT). For a rapid diagnosis and due to financial reasons, the owner agreed to advanced imaging, and temporarily declined the bile acid testing. CT examination under chemical restraint (Ketamine, Kepro, Deventer, The Netherlands, at 0.1 mg/kg IV and medetomidine, Domitor, Orion Pharma, at 0.05 mg/kg IV) revealed a large extrahepatic porto-caval shunt (Figure 2) associated with multiple conical cortical brain lesions that were hypodense, with reduced or absent contrast uptake, along with more extensive biconvex lesions, including along the cerebral sulci (Figure 3). A final diagnosis of portosystemic shunt and presumed hepatic encephalopathy was established.

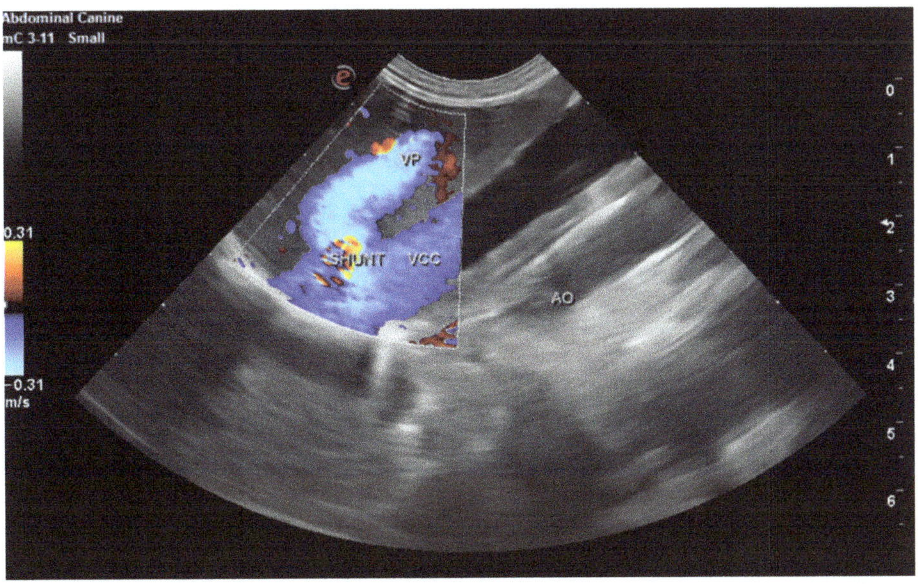

Figure 1. Abdominal ultrasound exam, with color Doppler reveal portocaval shunt. VP = portal vein. VCC = caudal cava vein. AO = abdominal aorta.

Following the CT study, under the same anesthesia, the bioelectrical activity of the brain was recorded using a Neurofax electroencephalograph (Nihon Kohden). Nine needle-type electrodes were placed as follows: F3 (left frontal), F4 (right frontal), C3 (left center), C4 (right center), O1 (left occipital), O2 (right occipital), A1 (left ear), A2 (right ear), and the ground Z electrode (positioned on the muzzle above the nose). Recordings were made in both a referential montage (F3-A1, C3-A1, O1-A1, F4-A2, C4-A2, and O2-A2) and a bipolar one (F3-C3, C3-O1, F4-C4, C4-O2, F3-F4, C3-C4, and O1-O2). The electroencephalographic recording was conducted using the following parameters: a sensitivity of 75 μV, a time constant of 0.3 s, a filter pass-down of 70 Hz, a filter pass-up of 30 Hz, and electrode impedance <10 Ω. The electrical activity of the brain was characterized by almost normal waves, represented by slow, symmetrical, and predominantly delta-wave background activity (Figure 4). The delta rhythm is commonly found in deep sleep or during anesthesia. In the case of anesthesia, once its effect ends, background activity is characterized by the appearance of a high-frequency waves that gradually pass through theta rhythms (Figure 5) to alpha and beta rhythms, with the latter marking the onset of waking status.

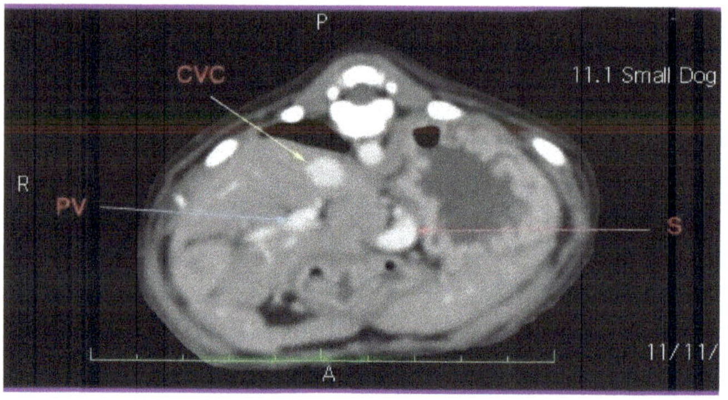

Figure 2. Abdominal CT exam. Large extrahepatic portocaval shunt, Bichon dog, 3 months. Following the caudal vena cava (CVC) from the caudocranial direction, it is observed that it runs parallel to the portal vein to the prerenal level (cranial to the kidneys), where it gives a ventral branch that descends ventro-laterally to the left, follows the small curvature of the stomach, then turns to the right and joins the portal vein (PV) at the level of the hepatic hilum.

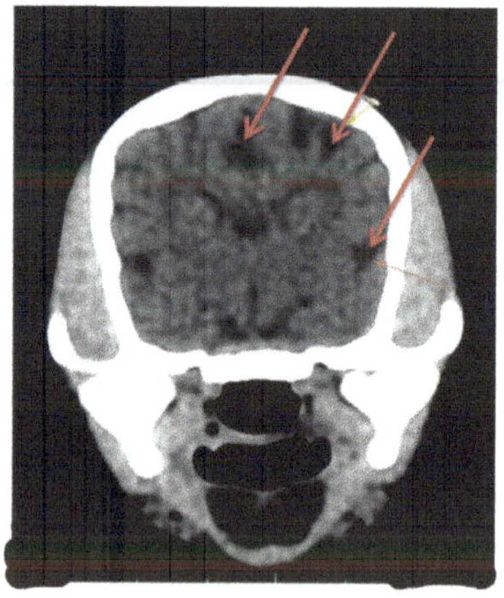

Figure 3. Cerebral CT exam in a Bichon dog, 3 months old, with portocaval shunt. Multiple hypoattenuated lesions with little to no contrast enhancement, some located in the cortex with a conical shape, others more extensive and biconvex, as well as along the cerebral sulci.

In the case of the described patient, a few minutes after the onset of the theta rhythm, the appearance of periodic, rhythmic, low-frequency discharges and a very high amplitude (over 150 microvolts) was observed on the EEG traces. Graphoelements could be identified in all leads, both in mono- and bipolar montages in a synchronous clinical context, represented by the installation of pedaling movements and rhythmic vocalizations (Figure 6). In addition to the hypersynchronized cerebral electrical activity, "bilateral triphasic waves" (Figure 7) were identified. These were characterized by complexes with a moderate to high amplitude (100–300 μV) of a frequency of 1.5 to 5 Hz (Figure 8). Although they were frequently prevalent in the frontal regions, they could also be seen in the occipital channels. The initial negative component was the sharpest, while the next positive part of the complex was the largest and subsequently followed by another negative wave. Persistent asymmetry (that is not related to technical factors or a cranial defect) may suggest an underlying structural injury on the side with the lower amplitude.

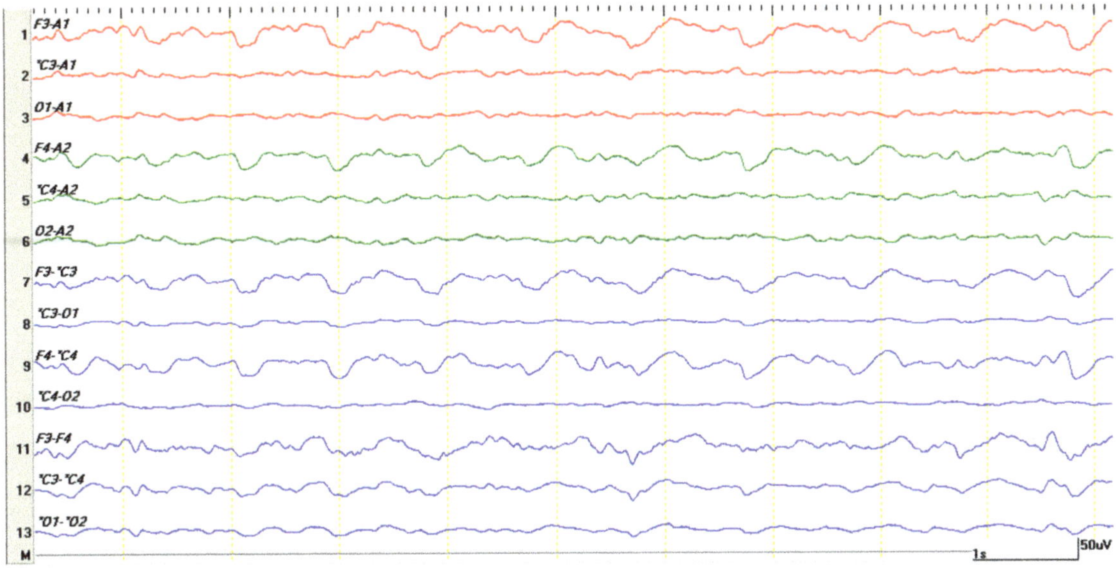

Figure 4. EEG recording for 10 s. The nearly normal cerebral activity of the patient immediately after the CT study, while still under medetomidine–ketamine anesthesia, showed slow, symmetric background activity characterized by delta waves (with a frequency of 1–4 Hz). Each second of the recording is delimited by the yellow vertical markings.

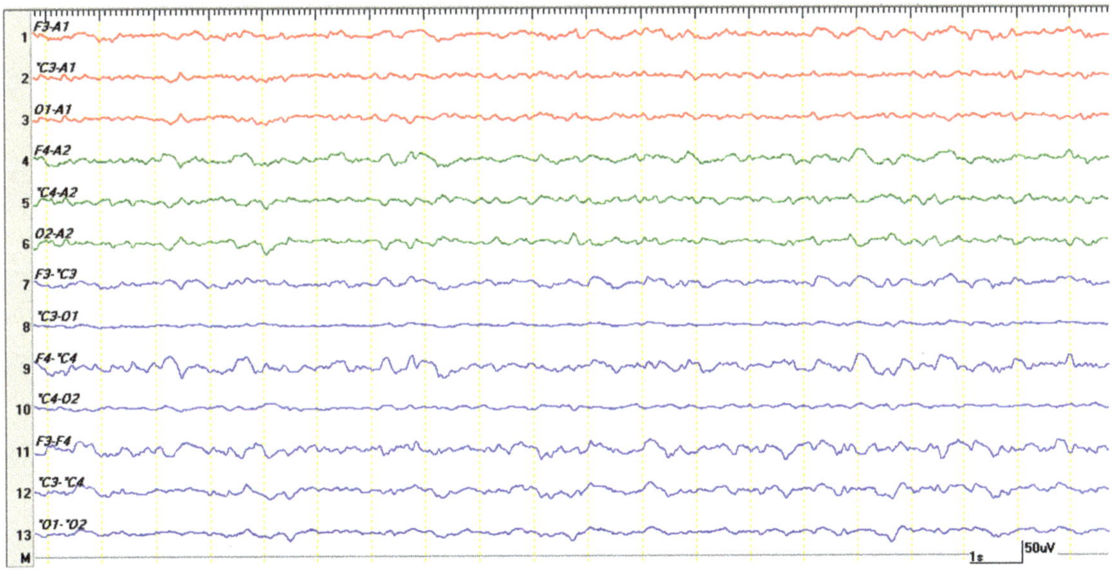

Figure 5. Slow background rhythm, theta (same patient) appears as the effects of anesthesia from the CT examination wear off. Compared to the previous image, a gradual increase in frequency and amplitude can be observed.

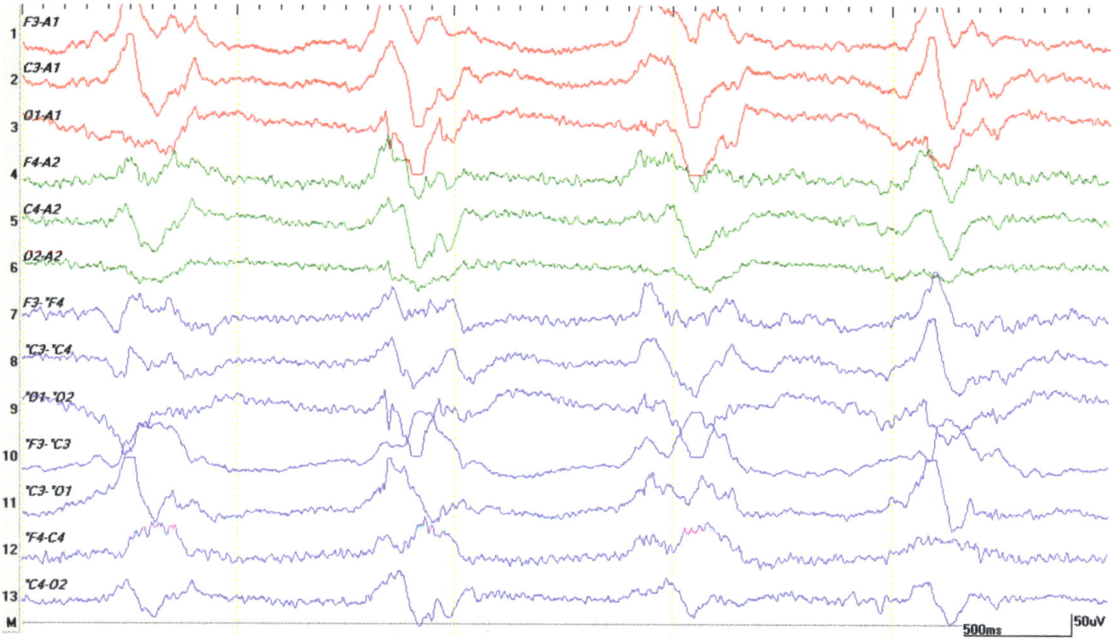

Figure 6. Periodic discharges with low frequency but high amplitude, reaching and exceeding 150 microvolts. Clinically, the patient exhibited paddling movements and rhythmic vocalizations synchronized with the graphoelements.

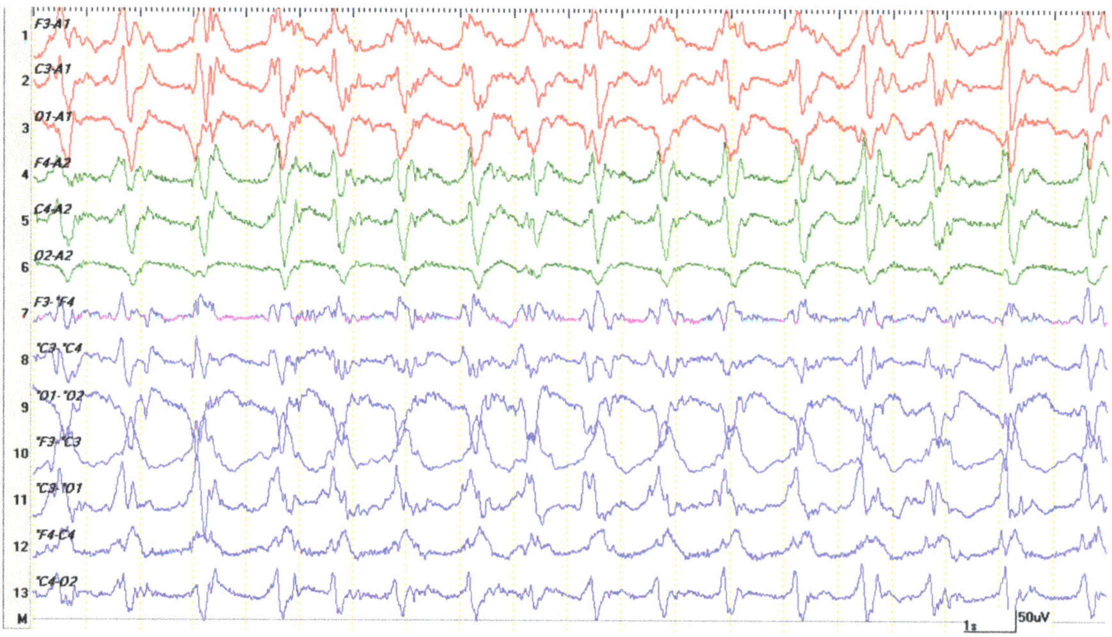

Figure 7. This figure displays a 20 s segment, compressed from the previous EEG trace, revealing the hypersynchronization of cerebral electrical activity. The primary graphoelements, which may resemble spikes, consist of sharp and slow wave complexes with three distinct phases, known as "triphasic waves".

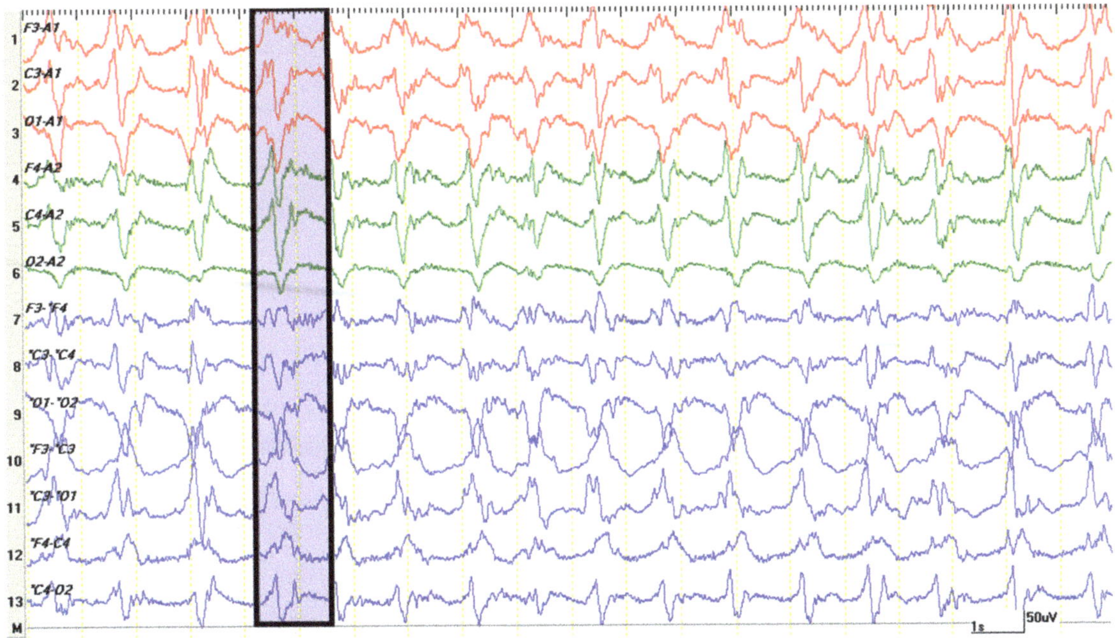

Figure 8. The morphology of bilateral symmetric triphasic waves is characterized by a high amplitude (exceeding 70 µV) and consists of three distinct phases: an initial small negative deflection, a prominent positive deflection, and a prolonged, slow negative deflection that gradually increases in amplitude. These waves exhibit a diffuse and bilaterally synchronous distribution.

After presenting the diagnosis and treatment options, the owners opted for euthanasia.

3. Discussion

In veterinary practice, EEG is commonly used for confirming the diagnostic of idiopathic epilepsy (Tier III according to the International Veterinary Epilepsy Taskforce consensus [13]). Over the last few years, the availability of EEG equipment increased and a large number of studies covering both the technical standardization and the findings in different encephalopathies were published [15–18]. According to Luca et al. (2023), other EEG indications besides diagnosing epilepsy, identification of epileptic foci, and antiepileptic drug efficacy include determining brain death, sleep disorders, research purposes, and post-op brain surgery monitoring [19]. In this paper, we describe the EEG findings in a pediatric dog with presumed HE due to an extrahepatic portosystemic shunt.

Portosystemic shunt (PS) is one of the most common congenital liver diseases in dogs [20]. Clinical signs associated with PS in dogs are various and involve neurological, digestive, and urinary system. Neurological signs are the expression of the hepatic encephalopathy (HE), and may vary for very subtle one (i.e., apathy) to severe ones (seizure, coma) [2]. Despite the fact that PS is easily diagnosed when the blood tests and imaging results are performed, HE diagnosis may be challenging for the veterinarian. Several studies performed both in humans and dogs show that EEG may be considered an useful diagnostic tool in diagnosing (and evaluation) of HE, as long as it can be associated with the presence of bilateral symmetric triphasic waves, a characteristic bioelectric pattern. The morphology of bilateral symmetric triphasic waves is characterized by a high amplitude (>70 µV) and consisting of an initial small negative deflection, followed by a large positive deflection and a long, slow, extended negative deflection that gradually increases, with a diffuse and bilaterally synchronous distribution [21–23]. In our dog, immediately after the diagnostic of PS was fixed by both ultrasound examination and CT exam, under the same anesthesia, the EEG traces show the presence of bilateral symmetric triphasic waves.

In humans, other potential causes for the occurrence of bilateral symmetric triphasic waves include focal structural lesions and concomitant toxic or metabolic encephalopathy (hyperthyroidism, hypo/hyperglycemia, hypo/hypercalcemia, thyamine deficiency, etc.) [14,21,22]. In veterinary patients, the presence of EEG periodic discharges was observed beside HE secondarily to PS in seizuring dogs due to a wide variety of etiologies, both intracranial and systemic, during the preictal, ictal or postictal period making their interpretation challenging [15]. Specifically, according to the same research group, triphasic waves were noticed in relationship with the presence of intra-axial mases, polycythemia and idiopathic epilepsy [15]. In our dog, based on the clinical symptomatology, blood tests and imaging results, we suspect that the occurrence of the bilateral triphasic waves were directly related with the development of HE. In a study conducted on 12 dogs with hepatic encephalopathy (due to portosystemic shunt) the EEG recordings revealed generalized high-voltage, low-frequency bioelectric activity, with the presence of symmetrical bilateral triphasic waves in 58% of cases [24]. In humans with HE, the presence of triphasic waves is a marker of severe HE (potentially as a result of white matter edema) and is a and indirect marker of increased risk of death [25].

In dogs, the occurrence of HE was historically associated with high ammonia levels, and some researchers state that the occurrence of triphasic waves in patients with HE is directly dependent on the levels of ammonia in systemic circulation [25]. However, both old and recent studies underline the idea that the HE may appear even in patients with a normal blood ammonia level. Moreover, in humans, the fasting ammonia concentration is nowadays considered an unreliable indicator of the degree of encephalopathy [26] and in dogs, the correlations between blood ammonia concentrations and disease severity (i.e., hepatic encephalopathy grade) were rather weak for dogs with extrahepatic portosystemic shunt [27]. Even though the absence of ammonia determination may be considered as a potential limitation in HE diagnosis in our dog, the clinical symptomatology and the imaging data were highly suggestive for a severe HE diagnostic.

At the time of presentation, the history (seizures) and cranial nerve deficits suggested a neurological localization at the level of the forebrain, with profound disruption of neuronal homeostasis in this area demonstrated by the electrical activity recorded in the EEG. However, it is interesting that, at its presentation, the patient exhibited decerebellate rigidity, suggesting a cerebellar lesion (and possibly an edema with "mass phenomenon"). The EEG technique does not record the activity of cerebellar neurons; therefore, we could not assume that an electrical alteration in the brain was necessarily present in the cerebellum as well. Furthermore, CT examination is not the preferred method for investigating cerebellar parenchyma in small animals. Magnetic resonance imaging (MRI) is a more suitable method for investigating the cerebellum, which represents a limitation in our study [28]. However, according to Weiss and Thabut [29], CT scans are still useful for exploring the pathophysiology of HE, whereas brain MRIs are cumbersome to perform in critically ill patients, and frequently not easily available. Also, the authors underline that for hepatic encephalopathy, CT scans could give some surrogate markers for evaluating treatment strategies [29]. In rats with experimentally induced portosystemic shunts, the morphological evidence demonstrated, in addition to HE, cerebellar impairment, including Purkinje cell loss, astrocyte and Bergmann glial cell swelling, microglial activation, and the onset of cytotoxic edema [30]. In dogs, congenital portosystemic shunting was associated with marked cellular changes in both cerebral cortex and cerebellum [31]. We suspect that the decerebellate rigidity may have been associated with cerebellar edema, which may have been of a higher grade than the compensated cerebral edema.

Another potential limitation of our study could be that total bile acid measurement were not performed. Bile acids test is a helpful marker in diagnosing the portosystemic shunt (and not the hepatic encephalopathy). The test's diagnostic sensitivity and specificity range from 88% to 93% and from 68% to 87%, respectively [32,33]. As a consequence, based only on bile acids test a diagnostic of portosystemic shunt (even if the test is a positive one) is only suspected but not confirmed. The advance imaging (as with computer tomography and Doppler ultrasonography) are preferred for establishing a final diagnosis [2]. In our manuscript, we proved the shunt by both imaging examinations: CT examination and Doppler ultrasonography. As a consequence, in this light, we consider that the existence of the shunt was proved by more accurate techniques (and not only suspected based on a potential bile acids test).

In the case of the presented patient, upon emergence from anesthesia, the EEG trace was characterized by continuous abnormal hypersynchronous activity, similar to non-convulsive status epilepticus (Figure 8). Triphasic waves are not epileptiform per se, and are not usually associated with seizures. However, generalized triphasic waves can resemble the pattern of non-convulsive status epilepticus. In the few studies conducted in human medicine, differentiating EEG traces between these two entities is difficult, and the fact that both resolve after the administration of diazepam may lead to diagnostic confusion [34,35]. In our case, although the patient had a history of seizures, the recorded EEG traces could not be definitively classified as non-convulsive status epilepticus. In the case of uncertainty, the application of an emergency anti-convulsant medication is preferred. However, despite benzodiazepine being considered the first line medication in treating status epilepticus in dogs [36], the use of benzodiazepine for stopping status epilepticus in dogs with HE due to PS may be controversial as long as no clinical trials exist in veterinary medicine, and in humans, their administration might precipitate the HE [2]. In such situation levetiracetam is considered to be a safer drug in achieving the therapeutic goal. In this light, in dogs with suspected HE secondarily to PS and triphasic waves on EEG recordings the use of EEG might be offer a rapid treatment orientation without increasing the outcome risks. As we previously mentioned, in humans, the triphasic waves pattern occurs especially in 2nd to 3rd grade HE [10]. Hence, repeated EEG recordings in dogs with HE may be useful for the objective assessment of the recovering the brain functionality and indirect the therapeutic efficacy of PS in dogs.

4. Conclusions

In this study, we describe a pediatric canine patient with reactive epileptic seizures following a presumed hepatic encephalopathy due to an extrahepatic portosystemic shunt. The EEG trace was characterized by the presence of bilateral synchronous three-phase wave bursts, resembling non-convulsive status epilepticus. EEG can complement the diagnosis of portosystemic shunting and provide an objective picture of the change in cerebral electrical homeostasis and might be useful in initiating the therapeutic protocol.

Supplementary Materials: The following supporting information can be downloaded at: https://www.mdpi.com/article/10.3390/life15010107/s1, Figure S1a. Abdominal ultrasound exam of the liver in healthy dog, M mode, reveal distinct traject of portal vein (VC) and caudal cava vein (VCC). Figure S1b. Abdominal ultrasound exam of the liver in healthy dog, Doppler mode, reveal distinct traject of portal vein (VC) and caudal cava vein (CVV).

Author Contributions: Conceptualization, R.A.Ș., M.D.C. and G.S.; writing—original draft preparation, R.A.Ș., V.B. and G.S.; writing—review and editing, R.A.Ș., M.M., M.D.C. and G.S.; supervision G.S. and M.D.C.; All authors have read and agreed to the published version of the manuscript.

Funding: This research received no external funding.

Institutional Review Board Statement: This study is a report on non-experimental clinical veterinary practice, part of clinical and veterinary diagnostics, and does not fall within Directive 63/2010 of the European Parliament and of the Council on the protection of animals used for scientific purposes. Ethical approval for the study was obtained from the Ethics Committee of the Faculty of Veterinary Medicine, University of Life Sciences "Ion Ionescu de la Brad" from Iasi (no. 454/14 March 2022). The animal in this study was examined and treated with the consent of its owners and handled in accordance with good veterinary practice.

Informed Consent Statement: This study is a report on non-experimental clinical veterinary practice, part of clinical and veterinary diagnostics, and does not fall within Directive 63/2010 of the European Parliament and of the Council on the protection of animals used for scientific purposes. The animal in this study was examined and treated with the consent of its owners and handled in accordance with good veterinary practice.

Data Availability Statement: The data presented in this study are available in the manuscript.

Acknowledgments: The authors thank Cristina Șerban for critical review of the CT exam.

Conflicts of Interest: The authors declare no conflicts of interest.

References

1. O'Brien, D. Toxic and Metabolic Causes of Seizures. *Clin. Tech. Small Anim. Pract.* **1998**, *13*, 159–166. [CrossRef] [PubMed]
2. Konstantinidis, A.O.; Patsikas, M.N.; Papazoglou, L.G.; Adamama-Moraitou, K.K. Congenital Portosystemic Shunts in Dogs and Cats: Classification, Pathophysiology, Clinical Presentation and Diagnosis. *Vet. Sci.* **2023**, *10*, 160. [CrossRef]
3. Lidbury, J.A.; Cook, A.K.; Steiner, J.M. Hepatic Encephalopathy in Dogs and Cats. *J. Vet. Emerg. Crit. Care* **2016**, *26*, 471–487. [CrossRef]
4. Gow, A.G. Hepatic Encephalopathy. *Vet. Clin. North Am. Small Anim. Pract.* **2017**, *47*, 585–599. [CrossRef] [PubMed]
5. Moon, S.-J.; Kim, J.-W.; Kang, B.-T.; Lim, C.-Y.; Park, H.-M. Magnetic Resonance Imaging Findings of Hepatic Encephalopathy in a Dog with a Portosystemic Shunt. *J. Vet. Med. Sci.* **2012**, *74*, 361–366. [CrossRef] [PubMed]
6. James, F.M.K.; Cortez, M.A.; Monteith, G.; Jokinen, T.S.; Sanders, S.; Wielaender, F.; Fischer, A.; Lohi, H. Diagnostic Utility of Wireless Video-Electroencephalography in Unsedated Dogs. *J. Vet. Intern. Med.* **2017**, *31*, 1469–1476. [CrossRef]
7. Stanciu, G.D. *Electrophysiological Exploration of the Nervous System in Veterinary Medicine*; Ion Ionescu de la Brad: Iași, Romania, 2016; ISBN 978-973-147-244-7. (In Romanian)
8. Wrzosek, M. Electroencephalography as a Diagnostic Technique for Canine Neurological Diseases. *J. Vet. Res.* **2016**, *60*, 181–187. [CrossRef]
9. Fisch, B.J. *Fisch and Spehlmann's EEG Primer: Basic Principles of Digital and Analog EEG*, 3rd ed.; Elsevier: Amsterdam, The Netherlands, 1999; ISBN 9780444821485.
10. Amodio, P.; Montagnese, S. Clinical Neurophysiology of Hepatic Encephalopathy. *J. Clin. Exp. Hepatol.* **2015**, *5*, S60–S68. [CrossRef]
11. Kullmann, F.; Hollerbach, S.; Lock, G.; Holstege, A.; Dierks, T.; Schölmerich, J. Brain Electrical Activity Mapping of EEG for the Diagnosis of (Sub)Clinical Hepatic Encephalopathy in Chronic Liver Disease. *Eur. J. Gastroenterol. Hepatol.* **2001**, *13*, 513–522. [CrossRef]
12. Montagnese, S.; Jackson, C.; Morgan, M.Y. Spatio-Temporal Decomposition of the Electroencephalogram in Patients with Cirrhosis. *J. Hepatol.* **2007**, *46*, 447–458. [CrossRef]
13. De Risio, L.; Bhatti, S.; Muñana, K.; Penderis, J.; Stein, V.; Tipold, A.; Berendt, M.; Farqhuar, R.; Fischer, A.; Long, S.; et al. International Veterinary Epilepsy Task Force Consensus Proposal: Diagnostic Approach to Epilepsy in Dogs. *BMC Vet. Res.* **2015**, *11*, 148. [CrossRef] [PubMed]
14. Faigle, R.; Sutter, R.; Kaplan, P.W. Electroencephalography of Encephalopathy in Patients with Endocrine and Metabolic Disorders. *J. Clin. Neurophysiol.* **2013**, *30*, 505–516. [CrossRef] [PubMed]
15. Knipe, M.F.; Bush, W.W.; Thomas, K.E.; Williams, D.C. Periodic Discharges in Veterinary Electroencephalography—A Visual Review. *Front. Vet. Sci.* **2023**, *10*, 1037404. [CrossRef] [PubMed]

16. Lyon, E.; Pochat, H.; Blot, S.; Troupel, T.; Van Caenegem, N.; Besnard, S.; Escriou, C. Use of Video-Electroencephalography as a First-Line Examination in Veterinary Neurology: Development and Standardization of Electroencephalography in Unsedated Dogs and Cats. *Front. Vet. Sci.* **2024**, *11*, 1326165. [CrossRef] [PubMed]
17. Musteata, M.; Ștefănescu, R.; Borcea, D.G.; Solcan, G. Very-Low-Frequency Spike-Wave Complex Partial Motor Seizure Mimicking Canine Idiopathic Head Tremor Syndrome in a Dog. *Vet. Sci.* **2023**, *10*, 472. [CrossRef] [PubMed]
18. Pellegrino, F.; Gómez Álvarez, C.M. Electroencephalographic Features of the Developing Brain in 72 Dogs under Xylazine Sedation: A Visual and Statistical Analysis. *Front. Vet. Sci.* **2023**, *10*, 1150617. [CrossRef]
19. Luca, J.; McCarthy, S.; Parmentier, T.; Hazenfratz, M.; Linden, A.Z.; Gaitero, L.; James, F.M.K. Survey of Electroencephalography Usage and Techniques for Dogs. *Front. Vet. Sci.* **2023**, *10*, 1198134. [CrossRef]
20. Watson, P. Canine Breed-Specific Hepatopathies. *Vet. Clin. North Am. Small Anim. Pract.* **2017**, *47*, 665–682. [CrossRef]
21. Fernández-Torre, J.L.; Kaplan, P.W. Triphasic Waves: Historical Overview of an Unresolved Mystery. *J. Clin. Neurophysiol.* **2021**, *38*, 399. [CrossRef]
22. Foreman, B. Can We Distinguish Triphasic Waves from Other Generalized Periodic Discharges? Do We Need To? *J. Clin. Neurophysiol.* **2021**, *38*, 362–365. [CrossRef]
23. Ștefănescu, R.A. Electroencephalographic Research in Epilepsy of Pet Carnivores (in Romanian). PhD Thesis, Iași University of Life Sciences "Ion Ionescu de la Brad", Iași, Romania, 2024.
24. Wrzosek, M.; Płonek, M.; Nicpoń, J. Electroencephalographic Features of Metabolic Encephalopathy in Dogs. *Med. Weter.* **2015**, *71*, 100–103.
25. Sutter, R.; Kaplan, P.W.; Cervenka, M.C.; Thakur, K.T.; Asemota, A.O.; Venkatesan, A.; Geocadin, R.G. Electroencephalography for Diagnosis and Prognosis of Acute Encephalitis. *Clin. Neurophysiol.* **2015**, *126*, 1524–1531. [CrossRef] [PubMed]
26. Rothuizen, J.; van den Ingh, T.S. Arterial and Venous Ammonia Concentrations in the Diagnosis of Canine Hepato-Encephalopathy. *Res. Vet. Sci.* **1982**, *33*, 17–21. [CrossRef] [PubMed]
27. Or, M.; Devriendt, N.; Kitshoff, A.M.; Peremans, K.; Vandermeulen, E.; Paepe, D.; Polis, I.; Martlé, V.; de Rooster, H. Ammonia Concentrations in Arterial Blood, Venous Blood, and Cerebrospinal Fluid of Dogs with and without Congenital Extrahepatic Portosystemic Shunts. *Am. J. Vet. Res.* **2017**, *78*, 1313–1318. [CrossRef] [PubMed]
28. Miguel-Garcés, M.; Gonçalves, R.; Quintana, R.; Álvarez, P.; Beckmann, K.M.; Alcoverro, E.; Moioli, M.; Ives, E.J.; Madden, M.; Gomes, S.A.; et al. Magnetic Resonance Imaging Pattern Recognition of Metabolic and Neurodegenerative Encephalopathies in Dogs and Cats. *Front. Vet. Sci.* **2024**, *11*, 1390971. [CrossRef]
29. Weiss, N.; Thabut, D. CT Scan Is Still a Valuable Tool to Assess Hepatic Encephalopathy Pathophysiology in Both Acute and Chronic Liver Diseases. *Crit. Care Med.* **2018**, *46*, e728. [CrossRef]
30. López-Cervantes, M.; Quintanar-Stephano, A.; Alcauter-Solórzano, S.; Hernández-Pando, R.; Aguilar-Roblero, R.; Gasca-Martínez, D.; Ortíz, J.J.; Vázquez-Martínez, O.; Ximénez-Camilli, C.; Díaz-Muñoz, M. Cerebellar Spongiform Degeneration Is Accompanied by Metabolic, Cellular, and Motor Disruption in Male Rats with Portacaval Anastomosis. *J. Neurosci. Res.* **2021**, *99*, 2287–2304. [CrossRef]
31. Williams, A.; Gow, A.; Kilpatrick, S.; Tivers, M.; Lipscomb, V.; Smith, K.; Day, M.O.; Jeffery, N.; Mellanby, R.J. Astrocyte Lesions in Cerebral Cortex and Cerebellum of Dogs with Congenital Ortosystemic Shunting. *J. Vet. Sci.* **2020**, *21*, e44. [CrossRef] [PubMed]
32. Gerritzen-Bruning, M.J.; van den Ingh, T.S.G.a.M.; Rothuizen, J. Diagnostic Value of Fasting Plasma Ammonia and Bile Acid Concentrations in the Identification of Portosystemic Shunting in Dogs. *J. Vet. Intern. Med.* **2006**, *20*, 13–19. [CrossRef]
33. Ruland, K.; Fischer, A.; Hartmann, K. Sensitivity and Specificity of Fasting Ammonia and Serum Bile Acids in the Diagnosis of Portosystemic Shunts in Dogs and Cats. *Vet. Clin. Pathol.* **2010**, *39*, 57–64. [CrossRef]
34. Boulanger, J.-M.; Deacon, C.; Lécuyer, D.; Gosselin, S.; Reiher, J. Triphasic Waves versus Nonconvulsive Status Epilepticus: EEG Distinction. *Can. J. Neurol. Sci.* **2006**, *33*, 175–180. [CrossRef]
35. Foreman, B.; Mahulikar, A.; Tadi, P.; Claassen, J.; Szaflarski, J.; Halford, J.J.; Dean, B.C.; Kaplan, P.W.; Hirsch, L.J.; LaRoche, S.; et al. Generalized Periodic Discharges and "Triphasic Waves": A Blinded Evaluation of Inter-Rater Agreement and Clinical Significance. *Clin. Neurophysiol.* **2016**, *127*, 1073–1080. [CrossRef]
36. Charalambous, M.; Muñana, K.; Patterson, E.E.; Platt, S.R.; Volk, H.A. ACVIM Consensus Statement on the Management of Status Epilepticus and Cluster Seizures in Dogs and Cats. *J. Vet. Intern. Med.* **2024**, *38*, 19–40. [CrossRef]

life

MDPI

Article

Evaluation of Biocomposite Cements for Bone Defect Repair in Rat Models

Alina Ioana Ardelean [1], Sorin Marian Mârza [2,*], Raluca Marica [3,*], Mădălina Florina Dragomir [1], Alina Oana Rusu-Moldovan [4], Mărioara Moldovan [5], Paula Maria Pașca [6] and Liviu Oana [1]

1 Department of Veterinary Surgery, Faculty of Veterinary Medicine, University of Agricultura Sciences and Veterinary Medicine, 3–5 Manastur Street, 400372 Cluj-Napoca, Romania; alina-ioana.ardelean@usamvcluj.ro (A.I.A.); madalina.dragomir@usamvcluj.ro (M.F.D.); oanaliviu2008@yahoo.com (L.O.)
2 Department of Veterinary Imagistics, Faculty of Veterinary Medicine, University of Agricultura Sciences and Veterinary Medicine, 3–5 Manastur Street, 400372 Cluj-Napoca, Romania
3 Department of Veterinary Pathology, Faculty of Veterinary Medicine, University of Agricultura Sciences and Veterinary Medicine, 3–5 Manastur Street, 400372 Cluj-Napoca, Romania
4 Department of Surgery III, Institute of Oncology "Prof. Dr. Alexandru Trestioreanu", 022328 Bucharest, Romania; alinaomoldovan@yahoo.com
5 Raluca Ripan Institute for Research in Chemistry, Babeș-Bolyai University, 30 Fantanele Street, 400294 Cluj-Napoca, Romania; marioara.moldovan@ubbcluj.ro
6 Clinics Department, Faculty of Veterinary Medicine, University of Agricultural Science and Veterinary Medicine, 700489 Iasi, Romania; paula.pasca@iuls.ro
* Correspondence: sorin.marza@usamvcluj.ro (S.M.M.); raluca.marica@usamvcluj.ro (R.M.)

Abstract: Repairing or reconstructing significant bone defects is typically challenging. In the present study, two composite cements were used as scaffolds in a sub-critical femoral defect in rats. A control group and two experimental batches were used to compare the outcomes. This research aimed to investigate the osteogenic potential and toxicological tolerance of the bioproducts through histopathology and computed tomography imaging analysis at 14, 28, 56, and 90 days post-implantation. The biomaterials used in the investigation consisted of a 65% bioactive salinized inorganic filler and a 25% weight organic matrix. The organic part of the biomaterial was composed of Bis-GMA (bisphenol A-glycidyl methacrylate), UDMA (urethane dimethacrylate), HEMA (2-Hydroxyethyl methacrylate), and TEGDMA (triethylene glycol dimethacrylate), while the inorganic filler was composed of silica, barium glass, hydroxyapatite, and fluor aluminosilicate glass. The first findings of this research are encouraging, revealing that there is a slight difference between the groups treated with biomaterials, but it might be an effective approach for managing bone abnormalities. Material C1 exhibited a faster bone defect healing time compared to material C2, where bone fractures occurred in some individuals. It is unclear if the fractures were caused by the presence of the biomaterial C2 or whether additional variables were to blame. By the end of the research, the mice appeared to tolerate the biomaterials without exhibiting any inflammatory or rejection responses.

Keywords: sub-critical bone defect repairment; rats; femur; biomaterial; composite cement scaffolds

1. Introduction

The body's bone structure has an extraordinary capability to regenerate and restore itself. The thickness and organization of skeletal trabeculae, morphology, shape, size, and especially the thickness of the cortical area all have a substantial impact on the physical and mechanical properties of the skeleton [1].

Bone abnormalities can arise as a result of trauma, neoplasia, or infection. While autologous grafts are considered the gold standard in managing such abnormalities, restricted accessibility and associated comorbidities limit their widespread use [2,3]. Synthetic skeletal–prosthetic polymers suggest promising results for the restoration of orthopedic defects [4]. In orthopedics, cement scaffolds are commonly utilized to replace skeletal grafts. The mineral structure of the cements is comparable to the composition of bone, which can create a powerful connection with the bone cells [5,6]. Bone cement is utilized in the musculoskeletal domain for filling and restoring abnormal trauma sites. They are frequently employed in different orthopedic and dentistry fixations [7,8]. Despite their exceptional performance, the relatively slow disintegration rate of these compounds limits their broader clinical use. Balancing the decomposition speed of polymers with new bone development is still an issue that needs to be addressed [9].

Because of its superior mechanical qualities, Bis-GMA is a resin commonly used in oral healthcare [10–13]. The material's elasticity, longevity, and polymerization shrinkage all increase when a UDMA monomer is used in its stead [14–17]. Given their important tissue interactions, they encourage angiogenesis and bone formation [18–20]. An additional field where cement has been used substantially is orthopedics. PMMA, or polymethyl methacrylate, is a synthetic substance that has no biological activity and is often utilized to secure implants. PMMA is mostly used in hip and knee replacements. Its functionality is based on a mechanical interlock between the prosthesis and the irregularly shaped bone. Bone cement implantation syndrome is frequently linked to this material [21]. Moreover, blood or saline irrigation of the bone can influence PMMA's mechanical characteristics and impede bone growth. The mechanical characteristics of PMMA are mainly impacted by the addition of hydroxyapatite (HAP), which has been observed to either increase or decrease its strength [22,23]. Despite the implant's biological compatibility, research has demonstrated that different implants can cause an immediate inflammatory reaction that occasionally results in systemic problems [24]. However, tight control over their structure can successfully limit toxicity and biocompatibility [25].

Based on its biological activity, hydroxyapatite (HAP), a naturally occurring material known as calcium apatite [26], is used in ceramics in medicine. It can exhibit bacteriostatic properties and can stimulate bone regeneration, increasing the survival rate of PMMA bonding. Barium promotes cell adhesion and proliferation as well as the synthesis of angiogenic agents [27].

During new bone formation, certain biomaterials break down within the organism. The byproducts of the decomposition from the material induce little to no tissue reaction, and in other situations, they encourage bone regeneration. The compounds are regularly absorbed or eliminated from cells, and therefore it is essential to pay attention to the organs that process waste [9].

Long bones are developed by endochondral ossification, which appears throughout growth or fracture repair. During this stage, osteogenic cells replace intermediate cartilage, resulting in new bone [28]. Among all bones, the femur tends to be frequently utilized considering its length, causing it to be more suitable for surgical approaches [29].

Unlike the majority of other organs, long bones have the potential to heal automatically with little to no scarring [30].

A 'critical size defect' in bone refers to a skeletal injury that will not heal without treatment. By conventional description, it is the smallest structural abnormality that will not fully repair within an animal's natural life [31,32]. Nonetheless, these essential size differences should be differentiated from abnormalities in which nonunion is induced by a disease-causing mechanism rather than size [33].

Our model evaluates a load-bearing sub-critical bone lesion. Critical or sub-critical size bone defects can arise due to a variety of pathologies, including acute trauma, malignancies, hereditary anomalies, and severe infection [34]. In prior investigations, the systemic and local tissular biocompatibility of the materials was established using in vitro [35] and in vivo cutaneous [36], subcutaneous, and intramuscular testing [37], and the composite cements exhibited osteo-inductive activity without triggering a foreign body reaction.

Various reasons have been suggested in the literature as potential causes of the decreased durability of biomaterials, including debonding, residual air bubbles, cement cracks, and porosity, known as external factors. A number of internal aspects may affect a material's characteristics: structure, dimensions, form, arrangement of the particles, ratio, and impurities (blood, bone fragments, fluids) [38–40]. A detailed description of the cements, including the manufacturing process, has been provided in a previous study [35]. It is well established that the mixing techniques employed significantly influence the quality of the bone cement [40].

The purpose of this study was to evaluate the composite cements in vivo using a certain research method for bone restoration.

Animal models are the foundation of preclinical translational technology advancements. The rats should have as low a morbidity or mortality ratio as possible to yield accurate data [34]. There are various animal models used for bone development, such as mice, rats, rabbits, dogs, pigs, sheep, and goats, but rodent models have been predominant in practically all investigations due to their dimensions, affordability, accessibility, and simplicity of handling [31,41,42]. Rats possess the majority of the main human bones and a similar skeletal structure compared to humans [43]. However, there are specific limitations due to Haversian remodeling [44].

The primary goal of this research was to provide a novel technique for a sub-critical bone defect on a rat's femoral mid-diaphysis to give suitable clinical relevance. We also assessed the potential of bone to biologically regenerate over four distinct periods with and without external intervention.

2. Materials and Methods

2.1. Preparation of Bone Biomaterials

The materials were produced at the Raluca Ripan Institute for Research in Chemistry (ICCRR) in Cluj-Napoca, Romania.

The biomaterials used in this investigation consisted of a 65% bioactive salinized inorganic filler and a 35% weight organic matrix. The organic part of the material was composed of Bis-GMA (ICCRR-UBB, Cluj-Napoca, Romania) ((2,2-bis[p-(2′-hydroxy-3′-metacryloxypropoxy)phenyl]-propane), UDMA (urethane dimethacrylate), HEMA (hydroxyethyl methacrylate), and TEGDMA (triethylene glycol dimethacrylate) (Sigma Aldrich in Darmstadt, Germany) while the inorganic filler was composed of silica, barium glass, hydroxyapatite, and fluor aluminosilicate glass (synthesized to ICCRR-UBB, Cluj-Napoca, Romania). The research led to the development of bioproducts C1 and C2 by distributing particles in the organic phase. The photopolymerization phase was initiated with a camphorquinone photoinitiator (CQ) (0.5% relative to the liquid mixture)/amine (1%) as the initiator/activator using an O-Star Curing Light lamp (Guilin Woodpecker Medical Instrument, Co., Ltd., Guilin, China) for 20 s. The composition of both products is illustrated in Figure 1.

The microstructural characteristics and elemental composition of the cements were thoroughly examined and documented in a previous investigation conducted at our research facility. Moreover, early in vitro research was conducted before creating in vivo experiments. The research of Ardelean et al. offers all the data needed [35].

Before insertion, the biomaterials were prepared to match the defect in a cuboidal shape which measured 4.00 mm in length, 2.00 mm in height, and 3 mm in width (Figure 2). The cements were sterilized on the day of surgery to prevent contamination.

The biomaterials were autoclaved at 105 °C for 10 min using a Trade Raypa Steam Sterilizer (R. Espinar, S.L., Spania, AE-75 Dry) to decontaminate them before the implantation.

2.2. Ethics Statement

This research study was carried out at the Faculty of Veterinary Medicine's Establishment for Breeding and Use of Laboratory Animals in Cluj-Napoca, Romania, after the specimens had been obtained from the Experimental Medicine Center at the University of Medicine and Pharmacy Iuliu Hatieganu.

C2

C1

Material category	Specific Product	Manufacturer	Function
bioactive salinized inorganic filler	BaF_2	UBB-ICCRR, Cluj-Napoca Romania	Refines radiopacity and mechanical strenght
bioactive salinized inorganic filler	Fluoroaluminosilicate	UBB-ICCRR, Cluj-Napoca, Romania	Auguments anticariogenic properties
organic matrix	Bis-GMA	UBB-ICCRR, Cluj-Napoca Romania	Elevates mechanical strenght
organic matrix	UDMA	Aldrich, Steinheim, Germany	Optimizes flexibility
organic matrix	TEGDMA	Aldrich, Steinheim, Germany	Reduces viscosity
organic matrix	HEMA	Aldrich, Steinheim, Germany	Adhesive enhancer
Photoinitiator	Camphorquinone	Aldrich, Steinheim, Germany	Initiates polymerization
bioactive salinized inorganic filler	HA	UBB-ICCRR, Cluj-Napoca Romania	Optimizes biocompatibility ,bioactivity, strenght and stability
bioactive salinized inorganic filler	Silica	UBB-ICCRR, Cluj-Napoca, Romania	Improves resistance
bioactive salinized inorganic filler	BaO	UBB-ICCRR, Cluj-Napoca Romania	Boosts adiopacity and mechanical strenght
bioactive salinized inorganic filler	Quartz	UBB-ICCRR, Cluj-Napoca, Romania	Optimizes mechanical strenght and provides thermal stability

Figure 1. The composition of experimental materials and the functions of the components.

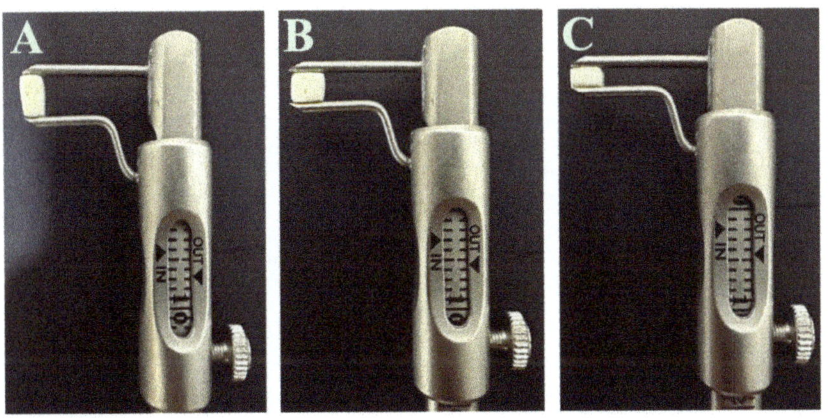

Figure 2. The representation of the biomaterial C1 in height (**A**), width (**B**), and length (**C**).

According to standards, the individuals used in the investigation received regular care, with consistent feeding intervals and closely monitored living conditions, such as a temperature of 23 °C, humidity cycles of 55%, and light/dark cycles of 12 h.

This study received approval from the Bioethics Committee of the University of Agricultural Sciences and Veterinary Medicine Cluj-Napoca no. 352/12.12.2022 and authorized by the Sanitary–Veterinary and Food Safety Department, Cluj-Napoca, through the Project Authorization no. 374/04.07.2023.

2.3. Blood Tests

Hematology and biochemistry investigations were conducted to confirm the rats' health condition. Blood was collected after the rats had been sedated, in specific blood-collecting tubes. A capillary tube was implanted at an angle of 30 degrees in the medial canthus of the eye [45]. The results were compared to specific literature [46,47].

2.3.1. Hematology Blood Test

Blood was processed with Abaxis VetScan HM5 hematology analyzer (Abaxis Inc., Union City, CA, USA). The full blood count was performed to measure the following: WBCs: White Blood Cells, LYMs: Lymphocytes, MONs: Monocytes, NEUs: Neutrophils, RBCs: Red Blood Cells, HCT: Hematocrit, HGB: Hemoglobin, MCV: Mean Cell Volume, MCH: Mean Corpuscular Hemoglobin, MCHC: Mean Corpuscular Hemoglobin Concentration, and PLT: Platelet count.

2.3.2. Biochemistry Blood Test

The blood was processed with an Automatic Veterinary Chemistry Analyzer Element RC (Scil Animal Care Company, Alfort, France). Several parameters were evaluated, such as ALB: Albumin, TP: Total Protein, TB: Total Bilirubin, ALT: Alanine Aminotransferase, ALP: Alkaline Phosphatase, CREA: Creatinine, UREA: Blood Urea Nitrogen, GLU: Glucose, CA: Calcium, PHOS: Phosphorus, K: Potassium, and NA: Sodium.

2.4. Animal Care and Use

For the animal resource in this investigation, we employed 24 adult female rats, divided into three groups, as follows: a control group, comprising 8 rats with no product (C0); group 1, composed of 8 rats with Cement 1 (C1); and group 2, consisting of 8 rats with Cement 2 (C2). The research rodent species were from the Muridae family, specifically the Wistar-Lewis line, around 350 g in weight and 10 weeks old. This variety was chosen because of its special attribute [48].

On days 0, 14, 28, 56, and 90, CT scans were performed. Blood samples were obtained on the first and last days of the research. Tissue samples were obtained for histological examination on days 14, 28, 56, and 90.

Two individuals per group were painlessly sacrificed through cervical dislocation while under general anesthesia, in conformity with international protocols, on each sacrificial day; the sacrificial days were days 14, 28, 56, and 90.

This study was carried out at the Faculty of Veterinary Medicine's Establishment for Breeding and Use of Laboratory Animals in Cluj-Napoca, Romania, after the specimens had been obtained from the Experimental Medicine Center at the University of Medicine and Pharmacy Iuliu Hatieganu.

2.5. Surgical Procedure

2.5.1. Anesthesiologic Protocol and Pain Control

To make the surgical process less unpleasant, we used an induction cage for the rodents. The anesthetic agent employed was isoflurane (Isothesia 250 mL, Omegavet, Bucuresti, Romania). The specimens remained in the box until the point at which they were unconscious, after which the intraperitoneal medication was administered. Rodents received anesthesia based on their body mass from a combination of ketamine (50 mg/kg Narkamon

Bio, Bioveta, Ivanovice na Hané, Czech Republic) and medetomidine (0.25 mg/kg Domitor, Biotur, Teleorman, Romania), injected intraperitoneally.

The rodents' eyes were protected against dryness using Corneregel eye gel (Bausch & Lomb, Rochester, NY, USA).

Following surgery, the rats received supplementary oxygen. Postoperatively, for analgesia we used Buprenorphine (1 mg/kg Bupaq, Biotur, Teleorman, Romania) subcutaneously in the first, second, and third days, one dose per 24 h.

2.5.2. Surgical Protocol

The animals were positioned in latero-lateral recumbency. From the sacral vertebrae to the tail and tarsal area, the right leg was clipped and aseptically prepared with diluted 4% Chlorhexidine and 70% Sanitary Alcohol. A 25 mm longitudinal incision was made with a #10 scalpel blade on the lateral side of the femur while holding it at full extension. The subcutaneous tissue was dilacerated with blunt dissection to expose the fascia of the Vastus Lateralis, Gluteal, and Biceps Femoris muscles. The dissection continued until the femur was exposed. Miniature Senn–Miller retractors were used to retract the muscular tissue, allowing for better visualization. Using an #11 scalpel blade, the periosteum was gently cut and lateralized out of the bone with a moisturized sterile Q-tip. The femur mid-diaphysis was then visible and prepared to create the bone defect.

Due to the modest size of the diaphysis, the bone defects were made using a neurosurgical 2 mm drill with a slow rotation speed (1500 rpm). During the defect creation, special care was given to the nutrient foramen to preserve it. The bone was cooled with a NaCl 0.9% (Chloride Sodium 0.9%, Braun, Ilfov, Romania) solution while drilling, using a 10 mL syringe and a 21 G needle to avoid overheating and cell apoptosis. Additionally, bone fragments and bone powder were carefully removed with this lavage to limit potential damage.

The defect's size was measured using a mechanical millimetric caliper. A cortical rounded-rectangular window 4.1 mm in length, 2.1 mm in height, and 3.1 mm in width (Figure 3A) was produced in the midsection of the diaphysis. The first group was the control group (C0). Their wound had the opportunity to heal naturally without any exterior interference. The second and third groups were provided with the C1 and C2 polymerized biomaterials (Figure 3B,C).

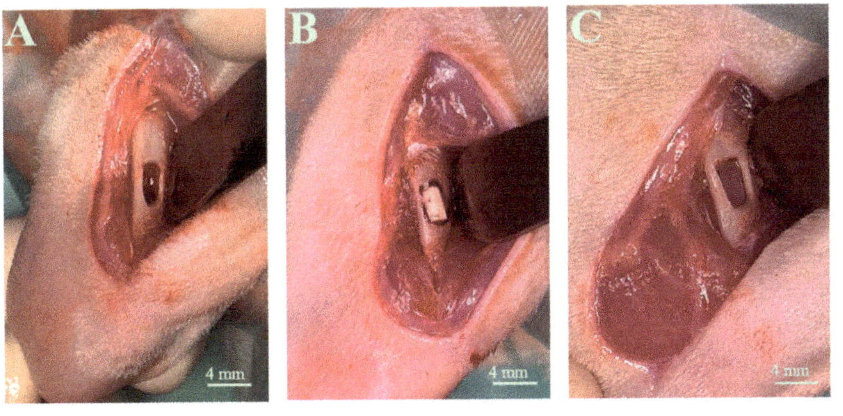

Figure 3. The bone defect was performed on the middle segment of the rats' femoral diaphysis of the right limb. Lateral exposure of the defect without (**A**) and with biomaterial C1 (**B**) and biomaterial C2 (**C**) on day 0.

The defect and surrounding tissue were lavaged with NaCl 0.9% solution before being closed in layers: muscle fascia with a simple interrupted pattern and subcutaneous tissue with a few buried knots using a 5.0 atraumatic absorbable polydioxanone monofilament

(PDO) (BioSintex, Snagov, Ilfov, Romania). The skin was closed using 4.0 nonabsorbable polyamide traumatic monofilament Nylon (BioSintex, BioNil Mono, Snagov, Ilfov, Romania) wire in a simple interrupted pattern.

2.6. Imaging

Starting on day one, all rats were scanned to observe the progression of the lesion. Aside from the progress, we also considered the possibility of fracture occurrence due to the destabilization of the femur.

Helical CT scanning of the right hind limb was obtained using a Siemens CT Somatom Scope machine with 16 channels. The scan was performed with the patient in sterno-abdominal recumbency. The patient was fully sedated using inhalation anesthetics (Isothesia, Baia-Mare, Romania).

Body images were obtained in the axial plane using a lower extremities protocol scan with a 512 × 512 matrix, narrow windows (WW: 120, WL: 40), 3 mm slice thickness, and a pitch of 3 mm, at KV 130 and mA 25. Multiplanar image reconstruction of the right hind limb was obtained using soft tissue and bone window reconstruction at a slice thickness of 0.6 mm.

The scans were performed on the day of implantation of the materials (day 0), and 14 days, 28 days, 56 days, and 90 days after implantation using the same protocol.

2.7. Histopathological Analysis of the Bones

Tissue specimens of the femur were collected at 14, 28, 56, and 90 days, and then preserved in 10% buffered neutral formaldehyde for 24 h. To protect the periosteum, the femur was extracted, and the surrounding muscles were carefully removed. This technique was used to isolate the femur for further histological examination while preserving the integrity of the periosteum and the bone. Histological sections from femoral defects were obtained directly from the site where the defects were created. Transverse planes were used to create the sections. To assess the effect of the biomaterials on the surrounding tissues and overall healing, the investigation encompassed both the primary defect area and peripheral areas. After fixation, the tissues underwent conventional histopathology processing. For examination, the paraffin-embedded samples were cut into 2-micrometer-thick slices and stained with Hematoxylin and Eosin (H&E) and Masson Trichrome (MT). Two separate pathologists evaluated the slides utilizing an Olympus BX40 microscope (Olympus Europa SE&Co, Hamburg, Germany). The photographs were captured with an Olympus SC 180 digital camera (Olympus Europa SE&Co, Hamburg, Germany) and prepared using Olympus CellSens, a specialized image acquisition and processing program.

On specific days, the samples were analyzed and contrasted. Both the C1 and C2 groups underwent a comparison with the Blank group.

3. Results

3.1. Biomaterials

A prior investigation thoroughly studied and reported on the microstructural features and the composition of the bioproducts. Furthermore, before the in vivo trials were created, early in vitro research was carried out. The research of Ardelean et al. provides all the information required [35].

After 90 days, the biomaterials remained unchanged and showed no signs of deterioration. Comparing the product to the compact bone with X H.U., the computed tomography scan revealed a greater radio-opacity. Furthermore, since the materials used possess a very high resistance to deformation or cutting, they were taken out of all samples before paraffin embedding.

3.2. Blood Tests

3.2.1. Hematological

The complete blood counts in all groups showed normal values equivalent to those in the control group (C0), exhibiting no statistically significant differences. The hematological profile data from days 1 through 90 of the study are displayed in Tables 1 and 2. The use of bioproducts did not negatively impact blood hematology.

Table 1. Mean and Standard Deviation of measured hematological parameters from the first day of the research. Statistics performed utilizing one-way ANOVA and two-way ANOVA [46,47].

		MEAN ± S.D.		
Analyte	**Blank**	**C1**	**C2**	**References**
WBC 10^9 cells/L	7.23 ± 0.84	7.36 ± 0.96	7.65 ± 0.73	2.10–19.50
LYM 10^9 cells/L	4.95 ± 0.37	5.03 ± 0.24	5.00 ± 0.42	2.00–14.10
MON 10^9 cells/L	0.41 ± 0.31	0.20 ± 0.09	0.51 ± 0.32	0.00–0.98
NEU 10^9 cells/L	1.98 ± 0.85	2.19 ± 0.96	2.36 ± 0.80	0.10–5.40
LYM %	68.57 ± 5.26	68.85 ± 6.40	67.28 ± 2.80	0–100
MON %	4.38 ± 1.00	3.85 ± 0.79	4.30 ± 1.20	0–100
NEU %	29.52 ± 7.70	27.43 ± 8.07	34.43 ± 3.34	0–100
RBC 10^{12} cells/L	8.11 ± 1.26	7.33 ± 0.73	8.86 ± 0.83	5.30–10
HGB g/dL	15.53 ± 2.42	14.01 ± 1.32	16.93 ± 1.70	14–18
HTC %	42.06 ± 5.42	38.99 ± 3.89	45.54 ± 2.77	35–52
MCV fl	52.23 ± 1.14	52.00 ± 0.71	52.10 ± 1.33	50–62
MCH pg	20.70 ± 1.34	20.15 ± 1.33	20.35 ± 1.49	16–23
MCHC g/dL	37.22 ± 1.36	36.80 ± 0.85	37.23 ± 1.57	31–40
PLT 10^9 cells/L	647.83 ± 25.39	659.75 ± 15.11	648.00 ± 30.63	500–1370

Table 2. Mean and Standard Deviation of measured hematological parameters from the last day of the research. Statistics performed utilizing one-way ANOVA and two-way ANOVA [46,47].

		MEAN ± S.D.		
Analyte	**Blank**	**C1**	**C2**	**References**
WBC 10^9 cells/L	6.88 ± 0.43	7.10 ± 0.81	7.27 ± 0.71	2.10–19.50
LYM 10^9 cells/L	5.20 ± 1.07	5.23 ± 0.48	5.08 ± 0.94	2.00–14.10
MON 10^9 cells/L	0.53 ± 0.32	0.29 ± 0.23	0.54 ± 0.31	0.00–0.98
NEU 10^9 cells/L	2.97 ± 0.96	2.55 ± 1.13	2.95 ± 0.89	0.10–5.40
LYM %	74.07 ± 5.17	70.48 ± 6.09	71.28 ± 2.36	0–100
MON %	4.83 ± 0.82	4.09 ± 0.66	4.65 ± 1.14	0–100
NEU %	33.87 ± 13.22	29.93 ± 9.87	38.06 ± 3.03	0–100
RBC 10^{12} cells/L	7.21 ± 0.76	7.03 ± 0.67	8.11 ± 1.15	5.30–10
HGB g/dL	15.88 ± 1.75	14.57 ± 1.61	17.01 ± 1.23	14–18
HTC %	42.06 ± 4.38	40.61 ± 4.80	44.91 ± 3.33	35–52
MCV fl	57.48 ± 1.12	54.63 ± 2.83	54.66 ± 2.88	50–62
MCH pg	21.45 ± 2.01	20.54 ± 1.91	20.53 ± 1.83	16–23
MCHC g/dL	36.68 ± 1.26	36.68 ± 1.24	36.95 ± 1.33	31–40
PLT 10^9 cells/L	734.17 ± 21.36	707.25 ± 15.11	702.75 ± 30.47	500–1370

3.2.2. Biochemical

The biochemical profile of the whole blood was analyzed on the first and last days of this study; the results are subsequently presented in Tables 3 and 4. The biochemical test findings showed no abnormalities of any kind, including liver or kidney impairment.

3.3. Surgical Procedure

A defect the size of the biomaterial was created in the cortical bone, which also impacted the medullar space. During the procedure, no complications occurred. A minor hemorrhage developed, but was maintained under control when a cold chloride solution was applied. Additionally, no signs of fracture were visible.

The entire procedure took an average of 25 min to execute, but after getting used to the method, the time was considerably shorter, saving up to 10 min.

Table 3. Mean and Standard Deviation of measured biochemical parameters from the first day of the research. Statistics performed utilizing one-way ANOVA and two-way ANOVA [46,47].

		MEAN ± S.D.		
Analyte	**Blank**	**C1**	**C2**	**References**
ALB g/dL	4.34 ± 0.29	4.46 ± 0.28	4.39 ± 0.34	4.1–5.4
TP g/dL	7.54 ± 0.61	7.82 ± 0.50	7.42 ± 0.64	6.4–8.5
TB mg/dL	0.07 ± 0.11	0.08 ± 0.13	0.10 ± 0.12	0.0–0.6
ALT U/L	30.85 ± 1.77	30.50 ± 2.06	32.03 ± 0.67	26–37
ALP U/L	112.07 ± 17.39	104.83 ± 16.93	123.35 ± 8.46	70–132
CREA mg/dL	0.92 ± 0.30	0.88 ± 0.29	1.03 ± 0.27	0.5–1.4
UREA mg/dL	38.08 ± 1.80	37.74 ± 1.86	39.03 ± 1.45	34.28–40.70
GLU mg/dL	131.15 ± 7.74	132.44 ± 8.17	138.14 ± 3.62	114–143
CA mg/dL	11.29 ± 0.31	11.31 ± 0.37	11.40 ± 0.22	10.5–13
PHOS mg/dL	9.90 ± 2.64	8.98 ± 2.80	11.75 ± 0.32	5–13
K mmol/L	6.60 ± 0.58	6.53 ± 0.68	6.84 ± 0.33	5.3–7.5
NA mmol/L	147.27 ± 2.36	149.30 ± 2.10	142.10 ± 2.85	143–150

Table 4. Mean and Standard Deviation of measured biochemical parameters from the last day of the research. Statistics performed utilizing one-way ANOVA and two-way ANOVA [46,47].

		MEAN ± S.D.		
Analyte	**Blank**	**C1**	**C2**	**References**
ALB g/dL	4.86 ± 0.42	4.98 ± 1.42	4.82 ± 0.39	4.1–5.4
TP g/dL	7.71 ± 0.48	7.50 ± 2.12	7.72 ± 0.48	6.4–8.5
TB mg/dL	0.25 ± 0.11	0.20 ± 0.13	0.16 ± 0.10	0.0–0.6
ALT U/L	34.10 ± 2.28	33.11 ± 2.45	34.03 ± 2.80	26–37
ALP U/L	111.82 ± 16.01	109.11 ± 12.76	106.16 ± 9.47	70–132
CREA mg/dL	1.25 ± 0.09	1.21 ± 0.38	1.26 ± 0.09	0.5–1.4
UREA mg/dL	38.64 ± 1.73	37.38 ± 1.93	38.76 ± 1.70	34.28–40.70
GLU mg/dL	126.37 ± 8.75	119.89 ± 8.42	115.77 ± 7.09	114–143
CA mg/dL	11.99 ± 0.42	11.94 ± 0.31	11.87 ± 0.64	10.5–13
PHOS mg/dL	9.78 ± 0.96	8.23 ± 1.93	7.62 ± 1.24	5–13
K mmol/L	6.99 ± 0.30	6.86 ± 1.55	7.06 ± 0.26	5.3–7.5
NA mmol/L	147.53 ± 1.51	148.35 ± 2.45	146.70 ± 2.19	143–150

3.4. *Imaging*

A 16-channel Siemens CT Somatom Scope was used to perform helical CT scanning of the whole body.

For the control rats, the bone defects were identified on the scan from day 0; it was 4.1 mm long on the bone axis, 3.1 mm deep, and 2.1 mm wide, and it involved the compact and medullary bone of the right femoral diaphysis. On the scan on day 14, the defect maintained its length and width and healing was observed from the depth, the depth being 2.8 mm; on the scan from day 28, bone healing was still evident and although the length and width of the defect remained the same, the depth of the defect decreased and was 1.6 mm; on the scan from day 56, the bone defect was no longer visible on the 3D reconstruction and from the axial compact perspective the bone was restored, the defect area showing only a slight decrease in radio-opacity; finally, at the scan on day 90, both the medullary and cortical bone were completely healed without signs of vicious callus or low radio-opacity (Figure 4).

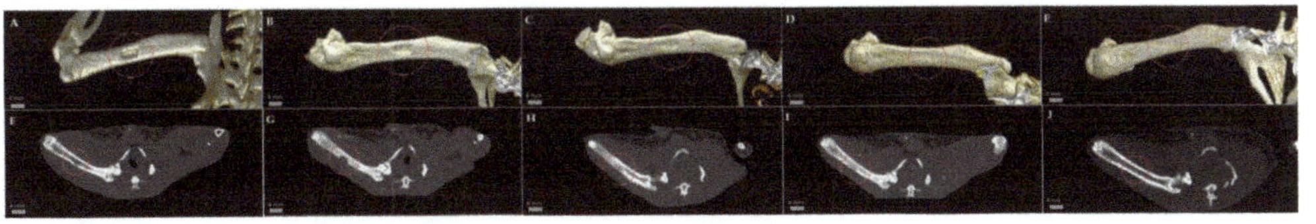

Figure 4. CT images showing the evolution of bone defect healing in the control rats. Three-dimensional reconstruction of the bone (**A–E**) and axial view (**F–J**): aspect from day 1 (**A**,**F**); aspect from day 14 (**B**,**G**); aspect from day 28 (**C**,**H**); aspect from day 56 (**D**,**I**); and aspect from day 90 (**E**,**J**); the area of the bone defect is marked with a red circle. Scale bar: 4 mm.

For the rats in which the C1 material was implanted on day 0, the material implanted at the level of the right femoral diaphysis was identified as having a length of 4.1 mm, a thickness of 3.1 mm, and a width of 2.1 mm and showed a higher radio-opacity than the compact bone with X H.U, and the muscle reaction zone at the level of the election site; at the scan on day 14, the implanted material was in the same position and of the same size, there was a minimal soft tissue reaction, and the compact bone was not healed; on the scan from day 28, the compact bone included the tested material with the exception of the proximal edge and the soft tissue no longer showed a reaction; on the scan from day 56, the compact bone covered the tested material with the exception of the proximal edge, and no defective callus reactions or soft tissue reactions were visible at that level; on the scan from day 90 in the area of the implant, the bony compact was completely healed and no defective callus reactions or soft tissue reactions were visible at that level (Figure 5).

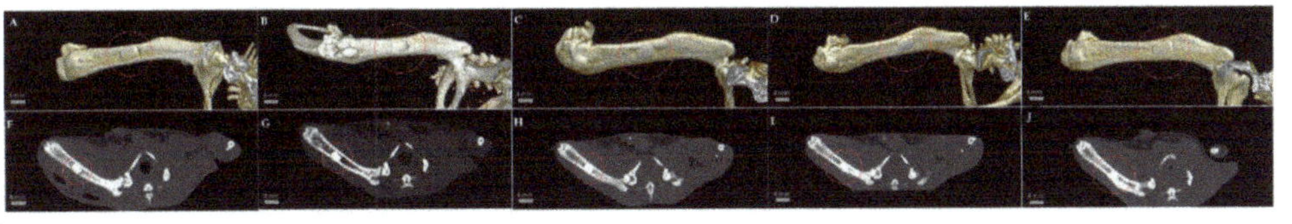

Figure 5. CT images showing the evolution of bone defect healing in the rats treated with the C1 material. Three-dimensional reconstruction of the bone (**A–E**) and axial view (**F–J**): aspect from day 1 (**A**,**F**); aspect from day 14 (**B**,**G**); aspect from day 28 (**C**,**H**); aspect from day 56 (**D**,**I**); and aspect from day 90 (**E**,**J**); the area of the bone defect is marked with a red circle. Scale bar: 4 mm.

For the rats in which the C2 material was implanted on day 0, the material implanted at the level of the right femoral diaphysis was identified as having a length of 4.1 mm, a thickness of 3.1 mm, and a width of 2.1 mm, and showed a higher radio-opacity compared to the compact bone with X H.U and the muscle reaction area at the level of the election site; at the scan on day 14, the implanted material was in the same position and of the same size and there was a minimal soft tissue reaction, the compact bone was not healed; on the scan from day 28, the compact bone incorporated the tested material and the soft tissue no longer showed a reaction; and on the scans from days 56 and 90, the compact bone covered the tested material and no defective callus reactions or soft tissue reaction were visible at that level (Figure 6).

In conclusion, from the point of view of CT scans, in rats treated with the C2 material, healing of the bone defect took place faster (28 days) compared to rats treated with the C1 material (56 days). No abnormal reactions in either soft tissue or bone tissue were observed in any of the materials.

3.5. Histopathological Examination

Prior to the paraffin embedding, the materials used to close the bone defect were removed from all samples, being too hard to section. In all the analyzed preparations, a cavity indicating a lack of substance was observed, corresponding to the material used. No phenomena of absorption or incorporation of the biomaterial were identified.

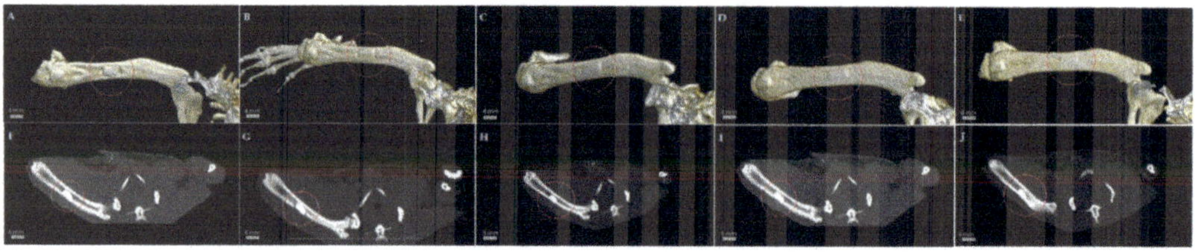

Figure 6. CT images showing the evolution of bone defect healing in the rats treated with the C2 material. Three-dimensional reconstruction of the bone (**A–E**) and axial view (**F–J**): aspect from day 1 (**A**,**F**); aspect from day 14 (**B**,**G**); aspect from day 28 (**C**,**H**); aspect from day 56 (**D**,**I**); aspect from day 90 (**E**,**J**); the area of the bone defect is marked with a red circle. Scale bar: 4 mm.

On day 14, the beginning of bone regeneration was observed in the C1 individuals; trabeculae of newly formed bone, immature woven bone, and gaps with hematogenous bone marrow could be identified. The inflammatory infiltrate was absent, and there was no fibrosis. A marked difference was noted in the C2 group, where, both at the level of the bone defect and the rest of the bone cortex, there were numerous collagen fibers, increased numbers of reactive fibroblasts, along with numerous blood vessels of different calibers (granulation tissue) and a diffuse inflammatory infiltrate dominated by polymorphonuclear cells. A focus of necrosis was present, with cellular and nuclear debris, a minimal interstitial hemorrhage, and the presence of fibrin and interstitial edema. Multifocal foci of cartilaginous metaplasia were identified; the transformation of fibrous tissue into mature cartilaginous tissue and the latter into newly formed bone lamellae (immature bone) were also observed.

A similar aspect was observed at 28 days, with the same biomaterial, where a traumatic bone fracture of the femur was noted with a significant loss of bone matrix, which extended beyond the edges of the bone defect. At the fracture site, a hematoma composed of erythrocytes mixed with fibrin was observed, surrounded by proliferative mesenchymal cells (callus) containing multiple variable blood vessels, fragments of bone tissue, collagen fibers, and rare multinucleated cells (osteoclasts). The bone tissue consisted of an increased number of osteoblasts, osteocytes, and rare osteoclasts. The subperiosteal bone tissue proliferates, composed of irregular and densely packed collagen fibers, formed a focus of hyaline cartilage adjacent to the fracture. It was unclear whether the biomaterials used caused the fractures in both individuals, or if other factors, such as the bone defect or the metabolic status of the individuals, were involved. In the C1 group, immature bone was identified, which extended from the subperiosteal level to the center, without fibrosis or inflammation.

At 56 days, the aspect of the control group was that of woven bone trabeculae, composed of dense collagen bundles and numerous osteocytes, similar to that identified in the C1 group. In the C2 group, a bone callus was identified, consisting of osteochondral trabeculae, varying in size, shape, and orientation. Dense fibrous connective tissue intersected with the trabeculae toward the periosteal surface, accompanied by a small number of plasma cells, lymphocytes, and hemosiderin-laden macrophages.

On the last day of the experiment, day 90, a complete regeneration of the bone tissue was observed in the control individual, with the created bone defect being completely closed. The width of the cortical bone was not identical along the entire circumference of the bone, being reduced at the level of the defect. In both the C1 and C2 groups, the cavity corresponding to the material used was completely delimited by newly formed bone tissue, composed of dense collagen fibers and numerous osteocytes. The medullary area was reduced in size, and we were able to observe multifocal islands of osteoid detached from the newly woven bone formed. Also, a complete regeneration of the periosteum was noted, which closed the bone defect, entrapping the biomaterial used.

Throughout the entire experiment (Figure 7), no necrotic or inflammatory changes were observed in the bone marrow for any of the materials used (Figure 8). Also, the biomaterials did not seem to excite any inflammatory or rejection response; they were well tolerated by the mice (Figure 9).

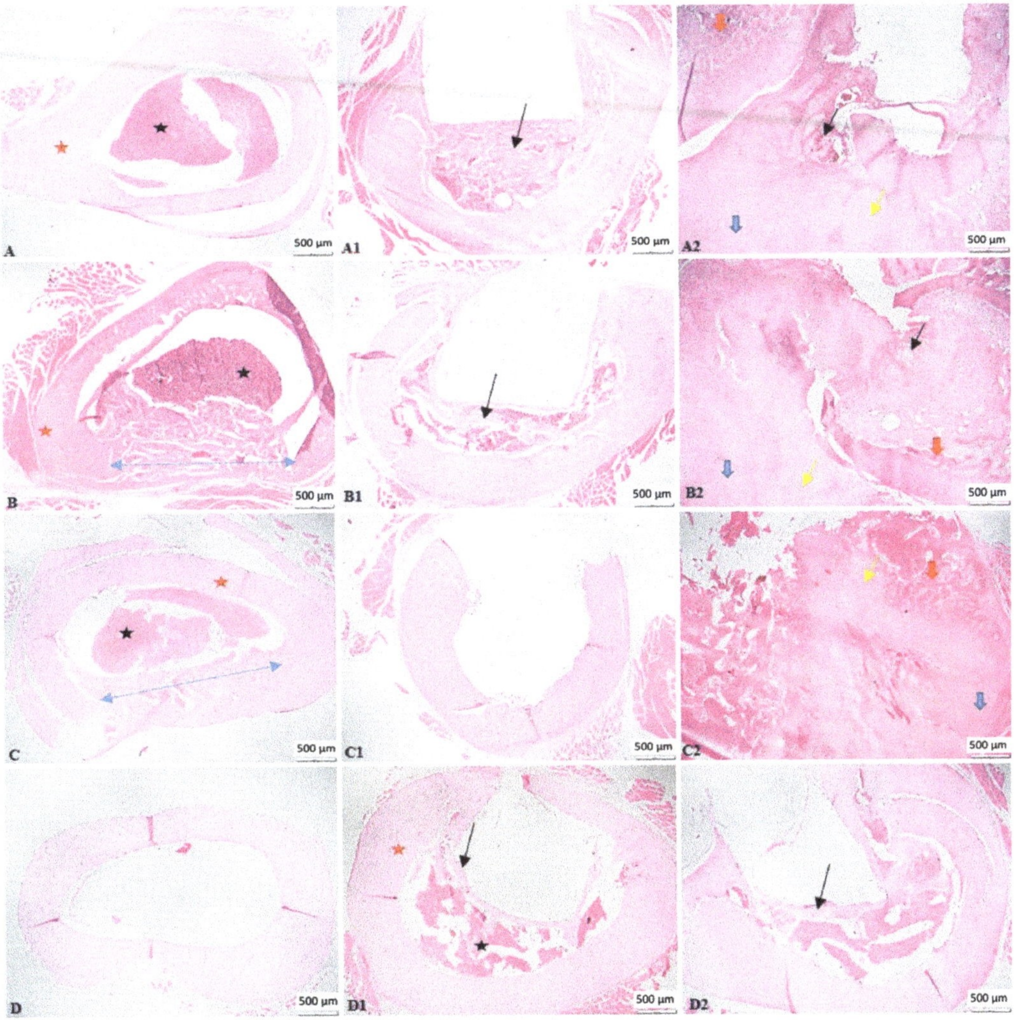

Figure 7. Photomicrographs of bone tissue—complete bone regeneration was observed in the control group; traumatic bone fracture was noted in all animals treated with (**C2**); by the end of the experiment, the cavity corresponding to the material used was completely delimited by newly formed bone tissue. (**A–D**) control group; (**A1–D1**) group treated with C1; (**A2–D2**) group treated with (**C2**). ((**A–C**)—blue arrow indicates bone defect); ((**A1**,**B1**,**D1**)—black arrow indicates woven bone); (black star indicates bone marrow; red star indicates compact bone); (**A2–C2**)—photomicrographs of callus, with blue arrows indicating collagen fibers, red arrows indicating bone trabeculae, black arrows indicating hematoma, and yellow arrows indicating cartilaginous metaplasia); (Duration: (**A–A2**) 14 days; (**B–B2**) 28 days; (**C–C2**) 56 days; (**D–D2**) 90 days). H&E, scale bar = 500 µm.

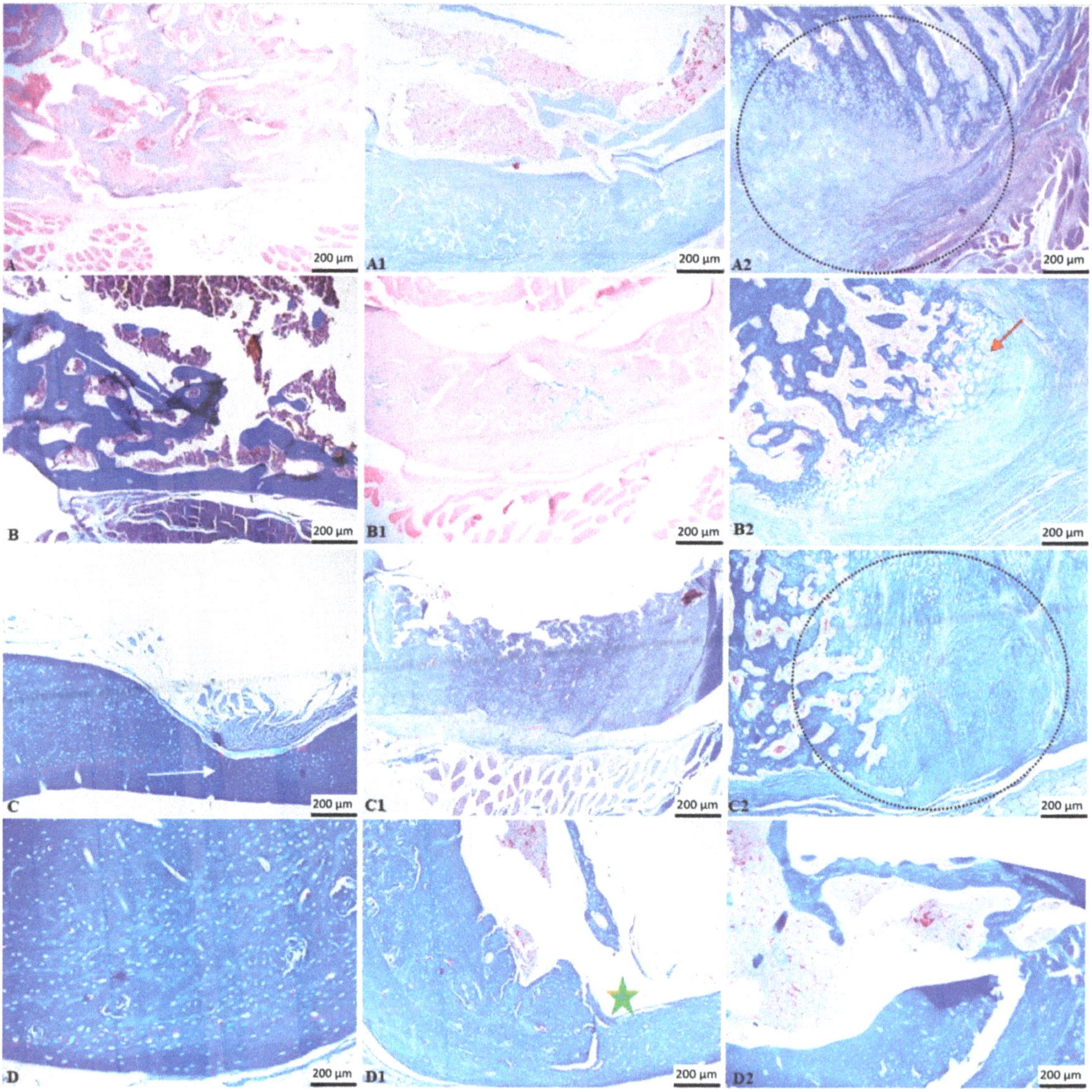

Figure 8. Photomicrographs of bone tissue. Bone regeneration was observed in the control group (white arrow); bone fracture and callus formation (delimited areas); cartilaginous metaplasia (red arrow) was noted in all animals treated with (**C2**); by the end of the experiment, the cavity corresponding to the material used was completely delimited by bone tissue (green star). (**A–D**) control group; (**A1–D1**) group treated with (**C1**); (**A2–D2**) group treated with (**C2**). Duration: (**A–A2**) 14 days; (**B–B2**) 28 days; (**C–C2**) 56 days; (**D–D2**) 90 days. TM, scale bar = 200 µm.

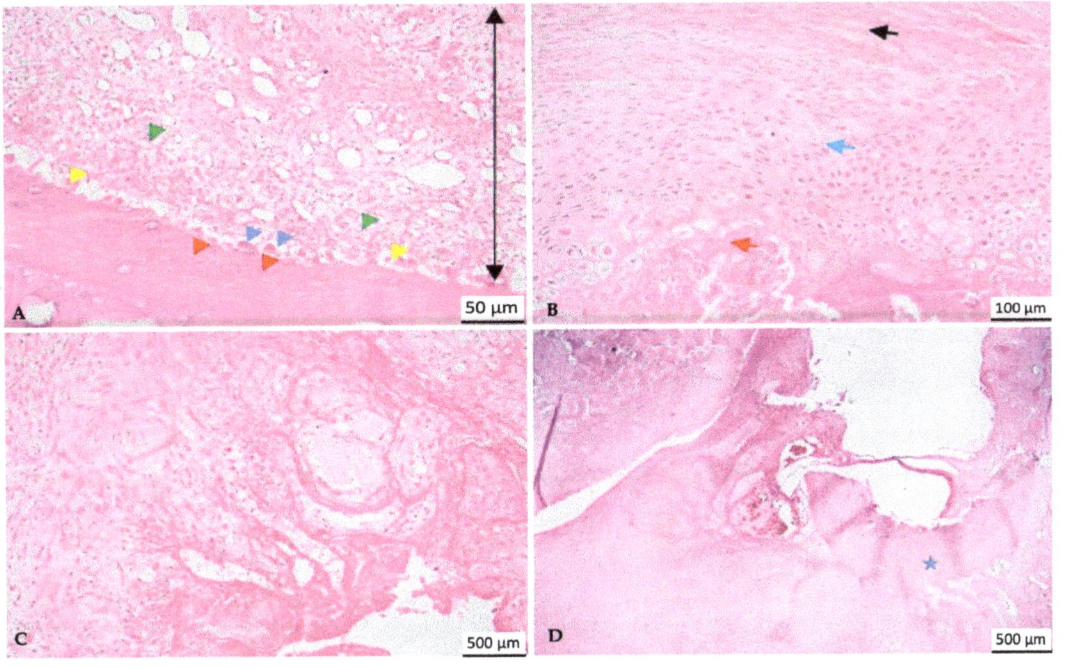

Figure 9. Detailed photomicrographs of bone tissue. (**A**)—granulation tissue (double arrow), containing multiple variable blood vessels and inflammatory cells (green arrowhead), accompanied by fragments of bone tissue, with rare osteoclasts (blue arrowhead), an increased number of osteoblasts (yellow arrow) and osteocytes (red arrowhead); H&E, Scale bar = 50 μm; (**B**)—endochondral ossification (fibrous tissue—black arrow; cartilaginous tissue—blue arrow; bone lamellae—red arrow); H&E, scale bar = 100 μm; (**C**)—hematoma with interstitial edema, fibrin, and granulation tissue; H&E, scale bar = 500 μm; (**D**)—fracture callus, surrounding compact bone tissue (star); H&E, scale bar = 500 μm.

4. Discussion

As potential options for more extensive bone tissue repair and support, biocompatible implants have garnered a lot of interest. Our products are a mixture of inorganic and organic components that, when combined, generate biocompatible biomaterials. This study assesses how low-molecular-weight monomers (Bis-GMA, TEGDMA, UDMA, and HEMA) interact with the filler fraction, which is composed of 65% silica, hydroxyapatite, barium glass, and fluor aluminosilicate glass. The experimental cementing materials C1 and C2 were produced by distributing the particles in the organic phase. Bis-GMA resin offers mechanical strength while UDMA increases elasticity and strength [49]. TEGDMA lowers viscosity to improve handling [50] and HEMA acts as an adhesion promoter, strengthening the adhesive's binding to the surrounding structures [51]. Excellent toughness and radio-opacity are features of barium glass [52]. At the same time, silica improves polishability and wear resistance [53].

Composites [45,46] for dentistry have been rapidly evolving over recent years, leading to their widespread usage as cosmetic tooth-like restorations [54,55]. We should not forget that dental composites might eventually be employed for bone augmentation, even though this application has not received as much interest. Critical-size bone deformities may potentially benefit from the usage of this type of bioproduct if they have been determined safe for use and biocompatible [56]. In this study, we investigated the osteo-inductive qualities, biocompatibility, and safety of composite cements when used as scaffolds in skeletal anomalies. The use of sub-critical defects can significantly affect the ability to assess the material's osteo-inductive capacity. Determining inductivity might be difficult in models when control defects are anticipated to heal completely over time. In this research, we can conclude that our model can reliably demonstrate that the biomaterial does not impair osteogenic functions.

We found that our bioproducts remained unchanged throughout the ninety-day investigation. Based on the products' chemical composition and the polymerization approach, neither biodegradation nor absorption of the products was observed. In a previous study, we investigated our products without polymerization on cutaneous defects, and the skin was able to absorb the biocomponents and heal effectively. During the polymerization process, the monomers create a strongly cross-linked system that is immune to decomposition by body fluids and enzymes. Fillers possess restricted dissolution qualities, reducing physiological disintegration, and are mechanically strong while remaining chemically inert. Prior research on hydroxyapatite has demonstrated that the addition of ceramic in the polymeric matrix affects the material's mechanical characteristics, speed of decomposition, and biological behavior in a dose-dependent manner [57–59].

Materials that fail to degrade properly can disrupt the natural bone healing process, potentially resulting in incomplete integration or prolonged complications. Non-degradable materials may also trigger chronic inflammation or adverse responses, ultimately jeopardizing the success of bone repair [7,8]. In this study, we examined the bone's reaction over a 90-day period and no side effects were seen. We acknowledge that the study duration is a limitation and future research should extend the observation period to achieve more comprehensive outcomes.

As previously mentioned, our polymerization approach preserves the mechanical strength of bone cements, beyond the mechanical limit of the bone. The majority of ceramic skeletal replacements are calcium-based and consist of a combination of tricalcium phosphate (TCP) and hydroxyapatite (HAP). Thus, it appears to become more and more likely to employ these composite cements scaffolds in living skeletal tissue.

The new bone starts chemical processes right after the surgery ends. First, the bone triggers a defense mechanism that attracts immune cells, and then the host mesenchymal cells migrate to the graft area through chemotaxis. Afterward, depending on several osteo-inductive signals, stem endogenous cells develop into chondroblasts and osteoblasts. Debris disintegration and revascularization take place subsequently. Ultimately, the restoration of bone takes place [60,61]. When put into the bone, the bioproducts have been observed to exhibit osteogenic capabilities. Nevertheless, the rate of resorption and disintegration is slower than that of spontaneous bone repair.

Accelerating the biomaterial's disintegration and enhancing the scaffold's resorption are two potential strategies that we will take into consideration for further studies [62–64]. However, one of this study's limitations is that bone healing failure may not be recognized until 15 weeks, which is three times longer than it takes for a normal fracture to heal [29]. It is important to note that the biomaterial C1 did not exhibit any fractures at the defect site, whereas the biomaterial C2 experienced fractures and elicited heightened healing responses. The cause of these fractures remains unclear, and further investigation is required to elucidate this issue. We consider the small number of rats used in the research a limitation, as a larger sample size would provide more robust data. Despite having all of these features, the products have proven to be biocompatible [65,66].

To demonstrate the biocompatibility of the composites, preliminary in vitro research on stem cells was conducted, followed by in vivo subcutaneous and intramuscular tests, before creating femoral defects. The bioproducts in both trials did not exhibit any evidence of rejection or cytotoxicity. Furthermore, in vitro tests demonstrated that the composite cements C1 and C2 display osteo-inductive behavior [35]. This ability can be evaluated through histopathological assessment.

Evaluating osteointegration is challenging because cement is highly dense, which may limit its infiltration capacity. Additionally, material removal during sectioning, although necessary due to the hardness of the cements, hinders the ability to accurately demonstrate cellular interaction. Yet, the bioproducts appear to be exhibiting the characteristics of osteoconductivity, as evidenced by the surrounding tissue's ability to close around it and its lack of resorption or retractions.

As was previously indicated, even though the defect was created in the cortical bone, it also affected the medullar area given the size of the biomaterials. In the bone marrow, mesenchymal stem populations originate from the medullary cavity. The destruction of the medullary space may impede the development of these precursor cells and thus affecting the bone's ability to heal [67]. None of these issues were discovered throughout our investigation. There were no indications of rejection, and the medullar space continued to generate progenitor cells.

Histopathology continues to be among the gold standards for evaluating biological compatibility for the clinical usage of bioproducts [68]. Bone segments were examined histopathologically at 14, 28, 56, and 90 days. Each specimen showed evidence of the bone defect. In all research subjects, tissue growth was evident; nevertheless, none of the biomaterials were absorbed after 90 days. During this day, the tissue allowed the biomaterial to integrate and interact. Individuals with trabeculae of newly formed bone, immature woven bone, and gaps where hematogenous bone marrow may be detected exhibited the onset of bone regeneration on day 14. On day 27, immature bone was visible, unaltered by fibrosis or inflammation and spreading from the subperiosteal level to the center. The characteristic observed at 56 days was woven bone trabeculae, which were made up of many osteocytes and thick collagen filaments. On day 90, newly generated bone tissue, consisting of many osteocytes and thick collagen fibers, totally delimited the cavity corresponding to the material employed. The size of the medullary region decreased, and multifocal osteoid islands that had separated from the freshly produced woven bone were noticed. Furthermore, the whole periosteum's regrowth was visible, sealing the bone defect and encasing the biomaterial. It is important to note that at the beginning of day 14, the C1 biomaterial provided superior results compared to the C2 biomaterial. Nevertheless, without causing a granulomatous reaction, the biomaterials demonstrated local biocompatibility. Despite requiring a lot of space, both composites encouraged positive biological effects and did not inhibit bone formation. Our subjects exhibited either no acute inflammatory cells or low levels of inflammation. A non-existent fibrous barrier surrounding the biomaterial and spontaneous regrowth of bone in proximity of the product was observed. There were no indications of necrosis or notable negative responses, which suggested positive biocompatibility.

The fact that the control in this model exhibited complete healing makes it challenging to compare these materials beyond evaluating their biocompatibility.

Rats were employed as a biological resource in this investigation. The ages of our individuals ranged from ten to sixteen weeks. Rats were selected based on their similarities with humans in terms of skeletal structure, reconstructive methods, and healing systems [69]. Even if tissue strength reduces with age, any age-related change may be minimal [70], since rats do not acquire complete ossification until they are one year old [71]. The study was conducted on the right femur to remove any further variations. All subjects were kept under the same conditions.

To validate the material's effectiveness and suitability for human clinical trials, future research should also assess its performance in larger animal models. All things considered, this work offers a solid platform for further studies and development of bioproducts-based treatments for bone reconstruction.

Since bone cement is a foreign component of the bone cement–prosthesis system, it is critical to comprehend the variables that might cause it to prematurely lose its mechanical qualities, which could cause the prosthesis to loosen. This feature, however, is considered a limitation of our study because it was not thoroughly examined in this research.

Some of the limitations of our study include the relatively small sample size of rats, the modest size of the bone defect, and the insufficient data on biomaterial degradation and mechanical strength. These constraints have made it difficult for us to obtain more thorough judgments about how effective bone cement is.

We intend to conduct a more thorough study to overcome these issues. Future studies will evaluate bone lesions of different sizes, include a bigger batch of rats, and collect more detailed information on mechanical characteristics and biomaterial degradation.

5. Conclusions

Osteocompatibility was demonstrated by both biomaterials.

The physico-chemical properties of the bioproducts fit the needs of the bone. CT scans confirmed the bioproducts' durability and mechanical strength.

The histological analysis showed no evidence of necrosis or rejection and instead supported the osteo-inductive feature of the biomaterials.

This could potentially be the starting point of a research project investigating the application of dental biomaterials in the production of scaffolds for bone defects.

Author Contributions: Conceptualization, A.I.A. and S.M.M.; methodology, M.M. and S.M.M.; software, A.I.A., A.O.R.-M. and M.F.D.; validation, M.M., S.M.M. and A.I.A.; formal analysis, A.O.R.-M., M.F.D. and R.M.; investigation, A.I.A., R.M. and P.M.P.; resources, R.M. and A.I.A.; data curation, A.I.A. and M.F.D.; writing—original draft preparation, A.I.A. and S.M.M.; writing—review and editing, M.M. and L.O.; visualization, A.I.A., L.O., M.M. and P.M.P.; supervision, M.M. and L.O. All authors have read and agreed to the published version of the manuscript.

Funding: This research received no external funding.

Institutional Review Board Statement: The experiment was approved by the Bioethics Committee of the University of Agricultural Sciences and Veterinary Medicine Cluj-Napoca no. 352/12.12.2022 and authorized by the Sanitary-Veterinary and Food Safety Department, Cluj-Napoca through the Project Authorization no. 374/04.07.2023.

Informed Consent Statement: Not applicable.

Data Availability Statement: Data are contained within the article.

Acknowledgments: Our deepest gratitude goes to Petrut Ghita for his invaluable support and assistance throughout the preparation of this paper. We sincerely appreciate his exceptional contributions and unwavering support.

Conflicts of Interest: The authors declare no conflicts of interest.

References

1. Olchowik, G.; Chadaj-Polberg, E.; Tomaszewski, M.; Polberg, M.; Tomaszewska, M. The Influence of Caffeine on the Biomechanical Properties of Bone Tissue during Pregnancy in a Population of Rats. *Folia Histochem. Cytobiol.* **2011**, *49*, 504–511. [CrossRef] [PubMed]
2. Giannoudis, P.V.; Dinopoulos, H.; Tsiridis, E. Bone Substitutes: An Update. *Injury* **2005**, *36* (Suppl. S3), S20–S27. [CrossRef] [PubMed]
3. Tu, Y.K.; Yen, C.Y.; Yeh, W.L.; Wang, I.C.; Wang, K.C.; Ueng, W.N. Reconstruction of Posttraumatic Long Bone Defect with Free Vascularized Bone Graft: Good Outcome in 48 Patients with 6 Years' Follow-Up. *Acta Orthop. Scand.* **2001**, *72*, 359–364. [CrossRef] [PubMed]
4. Jimenez-Marcos, C.; Mirza-Rosca, J.C.; Baltatu, M.S.; Vizureanu, P. Experimental Research on New Developed Titanium Alloys for Biomedical Applications. *Bioengineering* **2022**, *9*, 686. [CrossRef] [PubMed]
5. Heinemann, S.; Gelinsky, M.; Worch, H.; Hanke, T. Resorbable Bone Substitution Materials: An Overview of Commercially Available Materials and New Approaches in the Field of Composites. *Orthopade* **2011**, *40*, 761–773. [CrossRef]
6. Klein, A.; Baranowski, A.; Ritz, U.; Mack, C.; Götz, H.; Langendorf, E.; Al-Nawas, B.; Drees, P.; Rommens, P.M.; Hofmann, A. Effect of Bone Sialoprotein Coating on Progression of Bone Formation in a Femoral Defect Model in Rats. *Eur. J. Trauma Emerg. Surg.* **2020**, *46*, 277–286. [CrossRef]
7. Lodoso-Torrecilla, I.; van den Beucken, J.J.J.P.; Jansen, J.A. Calcium Phosphate Cements: Optimization toward Biodegradability. *Acta Biomater.* **2021**, *119*, 1–12. [CrossRef]
8. No, Y.J.; Xin, X.; Ramaswamy, Y.; Li, Y.; Roohaniesfahani, S.; Mustaffa, S.; Shi, J.; Jiang, X.; Zreiqat, H. Novel Injectable Strontium-Hardystonite Phosphate Cement for Cancellous Bone Filling Applications. *Mater. Sci. Eng. C Mater. Biol. Appl.* **2019**, *97*, 103–115. [CrossRef] [PubMed]

9. Liu, D.; Cui, C.; Chen, W.; Shi, J.; Li, B.; Chen, S. Biodegradable Cements for Bone Regeneration. *J. Funct. Biomater.* **2023**, *14*, 134. [CrossRef]
10. Luo, S.; Zhu, W.; Liu, F.; He, J. Preparation of a Bis-GMA-Free Dental Resin System with Synthesized Fluorinated Dimethacrylate Monomers. *Int. J. Mol. Sci.* **2016**, *17*, 2014. [CrossRef]
11. Ahovuo-Saloranta, A.; Forss, H.; Walsh, T.; Nordblad, A.; Mäkelä, M.; Worthington, H.V. Pit and Fissure Sealants for Preventing Dental Decay in Permanent Teeth. *Cochrane Database Syst. Rev.* **2017**, *7*, Cd001830. [CrossRef]
12. Haugen, H.J.; Marovic, D.; Par, M.; Khai, L.; Thieu, M.; Reseland, J.E.; Johnsen, G.F. Bulk Fill Composites Have Similar Performance to Conventional Dental Composites. *Int. J. Mol. Sci.* **2020**, *21*, 5136. [CrossRef] [PubMed]
13. Barszczewska-Rybarek, I.M.; Chrószcz, M.W.; Chladek, G. Novel Urethane-Dimethacrylate Monomers and Compositions for Use as Matrices in Dental Restorative Materials. *Int. J. Mol. Sci.* **2020**, *21*, 2644. [CrossRef]
14. Yoshinaga, K.; Yoshihara, K.; Yoshida, Y. Development of New Diacrylate Monomers as Substitutes for Bis-GMA and UDMA. *Dent. Mater.* **2021**, *37*, e391–e398. [CrossRef] [PubMed]
15. Szczesio-Wlodarczyk, A.; Domarecka, M.; Kopacz, K.; Sokolowski, J.; Bociong, K. An Evaluation of the Properties of Urethane Dimethacrylate-Based Dental Resins. *Materials* **2021**, *14*, 2727. [CrossRef]
16. Moszner, N.; Fischer, U.K.; Angermann, J.; Rheinberger, V. A Partially Aromatic Urethane Dimethacrylate as a New Substitute for Bis-GMA in Restorative Composites. *Dent. Mater.* **2008**, *24*, 694–699. [CrossRef] [PubMed]
17. Sarosi, C.; Moldovan, M.; Soanca, A.; Roman, A.; Gherman, T.; Trifoi, A.; Chisnoiu, A.M.; Cuc, S.; Filip, M.; Gheorghe, G.F.; et al. Effects of Monomer Composition of Urethane Methacrylate-Based Resins on the C=C Degree of Conversion, Residual Monomer Content and Mechanical Properties. *Polymers* **2021**, *13*, 4415. [CrossRef]
18. Hench, L.L. The Future of Bioactive Ceramics. *J. Mater. Sci. Mater. Med.* **2015**, *26*, 86. [CrossRef]
19. Crush, J.; Hussain, A.; Seah, K.T.M.; Khan, W.S. Bioactive Glass: Methods for Assessing Angiogenesis and Osteogenesis. *Front. Cell Dev. Biol.* **2021**, *9*, 643781. [CrossRef]
20. Ilie, N.; Erich Serfözö, N.; Prodan, D.; Diegelmann, J.; Moldovan, M. Synthesis and Performance of Experimental Resin-Based Dental Adhesives Reinforced with Functionalized Graphene and Hydroxyapatite Fillers. *Mater. Des.* **2022**, *221*, 110985. [CrossRef]
21. Al-Husinat, L.; Jouryyeh, B.; Al Sharie, S.; Al Modanat, Z.; Jurieh, A.; Al Hseinat, L.; Varrassi, G. Bone Cement and Its Anesthetic Complications: A Narrative Review. *J. Clin. Med.* **2023**, *12*, 2105. [CrossRef]
22. Karpiński, R.; Szabelski, J.; Krakowski, P.; Jonak, J.; Falkowicz, K.; Jojczuk, M.; Nogalski, A.; Przekora, A. Effect of Various Admixtures on Selected Mechanical Properties of Medium Viscosity Bone Cements: Part 1—α/β Tricalcium Phosphate (TCP). *Compos. Struct.* **2024**, *343*, 118306. [CrossRef]
23. Karpiński, R.; Szabelski, J.; Krakowski, P.; Jonak, J.; Falkowicz, K.; Jojczuk, M.; Nogalski, A.; Przekora, A. Effect of Various Admixtures on Selected Mechanical Properties of Medium Viscosity Bone Cements: Part 2—Hydroxyapatite. *Compos. Struct.* **2024**, *343*, 118308. [CrossRef]
24. Lupescu, O.; Nagea, M.; Scurtu, R.; Ciurea, N.M.; Dimitriu, A.L.; Marcov, N.; Popescu, G.I.; Bondari, S. Acute Cellulitis as Local Reaction to Orthopedic Implant—Case Presentation. *Rom. J. Morphol. Embryol.* **2016**, *57*, 1137–1143. [PubMed]
25. Roberts, T.T.; Rosenbaum, A.J. Bone Grafts, Bone Substitutes and Orthobiologics. *Organogenesis* **2012**, *8*, 114–124. [CrossRef]
26. Pokhrel, S. Hydroxyapatite: Preparation, Properties and Its Biomedical Applications. *Adv. Chem. Eng. Sci.* **2018**, *8*, 225–240. [CrossRef]
27. Majumdar, S.; Hira, S.K.; Tripathi, H.; Kumar, A.S.; Manna, P.P.; Singh, S.P.; Krishnamurthy, S. Synthesis and Characterization of Barium-Doped Bioactive Glass with Potential Anti-Inflammatory Activity. *Ceram. Int.* **2021**, *47*, 7143–7158. [CrossRef]
28. Mackie, E.J.; Ahmed, Y.A.; Tatarczuch, L.; Chen, K.-S.; Mirams, M. Endochondral Ossification: How Cartilage Is Converted into Bone in the Developing Skeleton. *Int. J. Biochem. Cell Biol.* **2008**, *40*, 46–62. [CrossRef]
29. Garcia, P.; Histing, T.; Holstein, J.H.; Klein, M.; Laschke, M.W.; Matthys, R.; Ignatius, A.; Wildemann, B.; Lienau, J.; Peters, A.; et al. Rodent Animal Models of Delayed Bone Healing and Non-Union Formation: A Comprehensive Review. *Eur. Cell Mater.* **2013**, *26*, 1–12, discussion 12–14. [CrossRef]
30. Evans, C.H. Advances in Regenerative Orthopedics. *Mayo Clin. Proc.* **2013**, *88*, 1323–1339. [CrossRef]
31. Schmitz, J.P.; Hollinger, J.O. The Critical Size Defect as an Experimental Model for Craniomandibulofacial Nonunions. *Clin. Orthop. Relat. Res.* **1986**, *205*, 299–308. [CrossRef]
32. Takagi, K.; Urist, M.R. The Reaction of the Dura to Bone Morphogenetic Protein (BMP) in Repair of Skull Defects. *Ann. Surg.* **1982**, *196*, 100–109. [CrossRef]
33. Einhorn, T.A. Clinically Applied Models of Bone Regeneration in Tissue Engineering Research. *Clin. Orthop.* **1999**, *367*, S59–S67. [CrossRef] [PubMed]
34. Spicer, P.P.; Kretlow, J.D.; Young, S.; Jansen, J.A.; Kasper, F.K.; Mikos, A.G. Evaluation of Bone Regeneration Using the Rat Critical Size Calvarial Defect. *Nat. Protoc.* **2012**, *7*, 1918–1929. [CrossRef]
35. Ardelean, A.I.; Dragomir, M.F.; Moldovan, M.; Sarosi, C.; Paltinean, G.A.; Pall, E.; Tudoran, L.B.; Petean, I.; Oana, L. In Vitro Study of Composite Cements on Mesenchymal Stem Cells of Palatal Origin. *Int. J. Mol. Sci.* **2023**, *24*, 10911. [CrossRef] [PubMed]
36. Ardelean, A.I.; Marza, S.M.; Dragomir, M.F.; Negoescu, A.; Sarosi, C.; Novac, C.S.; Pestean, C.; Moldovan, M.; Oana, L. The Potential of Composite Cements for Wound Healing in Rats. *Bioengineering* **2024**, *11*, 837. [CrossRef]
37. Ardelean, A.I.; Marza, S.M.; Negoescu, A.; Dragomir, M.F.; Sarosi, C.; Moldovan, M.; Ene, R.; Oana, L. Assessing Biocompatibility of Composite Cements by Peri/Intramuscular and Subcutaneous Implantation in Rats. *Biomedicines* **2024**, *12*, 1718. [CrossRef]

38. Karpiński, R.; Szabelski, J.; Maksymiuk, J. Seasoning Polymethyl Methacrylate (PMMA) Bone Cements with Incorrect Mix Ratio. *Materials* **2019**, *12*, 3073. [CrossRef]
39. Szabelski, J.; Karpiński, R.; Krakowski, P.; Jojczuk, M.; Jonak, J.; Nogalski, A. Analysis of the Effect of Component Ratio Imbalances on Selected Mechanical Properties of Seasoned, Medium Viscosity Bone Cements. *Materials* **2022**, *15*, 5577. [CrossRef]
40. Dunne, N.J.; Orr, J.F. Influence of Mixing Techniques on the Physical Properties of Acrylic Bone Cement. *Biomaterials* **2001**, *22*, 1819–1826. [CrossRef]
41. Cancedda, R.; Giannoni, P.; Mastrogiacomo, M. A Tissue Engineering Approach to Bone Repair in Large Animal Models and in Clinical Practice. *Biomaterials* **2007**, *28*, 4240–4250. [CrossRef]
42. Reichert, J.C.; Epari, D.R.; Wullschleger, M.E.; Saifzadeh, S.; Steck, R.; Lienau, J.; Sommerville, S.; Dickinson, I.C.; Schütz, M.A.; Duda, G.N.; et al. Establishment of a Preclinical Ovine Model for Tibial Segmental Bone Defect Repair by Applying Bone Tissue Engineering Strategies. *Tissue Eng. Part B Rev.* **2010**, *16*, 93–104. [CrossRef]
43. Bagi, C.M.; Berryman, E.; Moalli, M.R. Comparative Bone Anatomy of Commonly Used Laboratory Animals: Implications for Drug Discovery. *Comp. Med.* **2011**, *61*, 76–85. [PubMed]
44. Muschler, G.F.; Raut, V.P.; Patterson, T.E.; Wenke, J.C.; Hollinger, J.O. The Design and Use of Animal Models for Translational Research in Bone Tissue Engineering and Regenerative Medicine. *Tissue Eng. Part B Rev.* **2010**, *16*, 123–145. [CrossRef] [PubMed]
45. Hoff, J. Methods of Blood Collection in the Mouse. *Lab Anim.* **2000**, *29*, 47–53.
46. Kraft, W. *Clinical Laboratory Diagnostics in Veterinary Medicine*, 7th ed.; Schattauer GmbH: Stuttgart, Germany, 2014.
47. Vigneshwar, R.; Arivalagan, A.; Mekala, P.; Imayarasi, K. Sex-Specific Reference Intervals for Wistar Albino Rats: Hematology and Clinical Biochemistry. *Indian J. Anim. Health* **2021**, *60*, 58–65. [CrossRef]
48. Didion, J.P.; de Villena, F.P.-M. Deconstructing *Mus gemischus*: Advances in Understanding Ancestry, Structure, and Variation in the Genome of the Laboratory Mouse. *Mamm. Genome* **2013**, *24*, 1–20. [CrossRef]
49. Rosentritt, M.; Huber, C.; Strasser, T.; Schmid, A. Investigating the Mechanical and Optical Properties of Novel Urethandimethacrylate (UDMA) and Urethanmethacrylate (UMA) Based Rapid Prototyping Materials. *Dent. Mater.* **2021**, *37*, 1584–1591. [CrossRef]
50. Szczesio-Wlodarczyk, A.; Polikowski, A.; Krasowski, M.; Fronczek, M.; Sokolowski, J.; Bociong, K. The Influence of Low-Molecular-Weight Monomers (TEGDMA, HDDMA, HEMA) on the Properties of Selected Matrices and Composites Based on Bis-GMA and UDMA. *Materials* **2022**, *15*, 2649. [CrossRef]
51. Pimentel de Oliveira, R.; de Paula, B.L.; Ribeiro, M.E.; Alves, E.; Costi, H.T.; Silva, C. Evaluation of the Bond Strength of Self-Etching Adhesive Systems Containing HEMA and 10-MDP Monomers: Bond Strength of Adhesives Containing HEMA and 10 MDP. *Int. J. Dent.* **2022**, *2022*, 5756649. [CrossRef]
52. Rammah, Y.S.; El-Agawany, F.I.; Elkhoshkhany, N.; Elmasry, F.; Reben, M.; Grelowska, I.; Yousef, E. Physical, Optical, Thermal, and Gamma-Ray Shielding Features of Fluorotellurite Lithiumniobate Glasses: TeO_2-$LiNbO_3$-BaO-BaF_2-La_2O_3. *J. Mater. Sci. Mater. Electron.* **2021**, *32*, 3743–3752. [CrossRef]
53. Wang, R.; Bao, S.; Liu, F.; Jiang, X.; Zhang, Q.; Sun, B.; Zhu, M. Wear Behavior of Light-Cured Resin Composites with Bimodal Silica Nanostructures as Fillers. *Mater. Sci. Eng. C Mater. Biol. Appl.* **2013**, *33*, 4759–4766. [CrossRef]
54. Aminoroaya, A.; Neisiany, R.E.; Khorasani, S.N.; Panahi, P.; Das, O.; Madry, H.; Cucchiarini, M.; Ramakrishna, S. A Review of Dental Composites: Challenges, Chemistry Aspects, Filler Influences, and Future Insights. *Compos. Part B Eng.* **2021**, *216*, 108852. [CrossRef]
55. Dreanca, A.; Sarosi, C.; Parvu, A.E.; Blidaru, M.; Enacrachi, G.; Purdoiu, R.; Nagy, A.; Sevastre, B.; Oros, N.A.; Marcus, I.; et al. Systemic and Local Biocompatibility Assessment of Graphene Composite Dental Materials in Experimental Mandibular Bone Defect. *Materials* **2020**, *13*, 2511. [CrossRef] [PubMed]
56. Dreanca, A.; Bogdan, S.; Popescu, A.; Sand, D.; Pall, E.; Astilean, A.N.; Pestean, C.; Toma, C.; Marza, S.; Taulescu, M.; et al. The Evaluation of the Osteopromoting Capabilities of Composites Based on Biopolymers and Gold/Silver Nanoparticles Doped Bioactive Glasses on an Experimental Rat Bone Defect. *Biomed. Mater.* **2023**, *18*, 055014. [CrossRef] [PubMed]
57. Lin, P.-L.; Fang, H.-W.; Tseng, T.; Lee, W.-H. Effects of Hydroxyapatite Dosage on Mechanical and Biological Behaviors of Polylactic Acid Composite Materials. *Mater. Lett.* **2007**, *61*, 3009–3013. [CrossRef]
58. Barbieri, D.; Renard, A.J.S.; de Bruijn, J.D.; Yuan, H. Heterotopic Bone Formation by Nano-Apatite Containing Poly(D,L-Lactide) Composites. *Eur. Cell. Mater.* **2010**, *19*, 252–261. [CrossRef] [PubMed]
59. Barbieri, D.; de Bruijn, J.D.; Luo, X.; Farè, S.; Grijpma, D.W.; Yuan, H. Controlling Dynamic Mechanical Properties and Degradation of Composites for Bone Regeneration by Means of Filler Content. *J. Mech. Behav. Biomed. Mater.* **2013**, *20*, 162–172. [CrossRef]
60. Goldberg, V.M.; Stevenson, S. The Biology of Bone Grafts. *Semin. Arthroplast.* **1993**, *4*, 58–63.
61. Zipfel, G.J.; Guiot, B.H.; Fessler, R.G. Bone Grafting. *Neurosurg. Focus* **2003**, *14*, e8. [CrossRef]
62. Trincă, L.C.; Burtan, L.; Mareci, D.; Fernández-Pérez, B.M.; Stoleriu, I.; Stanciu, T.; Stanciu, S.; Solcan, C.; Izquierdo, J.; Souto, R.M. Evaluation of In Vitro Corrosion Resistance and In Vivo Osseointegration Properties of a FeMnSiCa Alloy as Potential Degradable Implant Biomaterial. *Mater. Sci. Eng. C Mater. Biol. Appl.* **2021**, *118*, 111436. [CrossRef] [PubMed]
63. Trincă, L.C.; Mareci, D.; Solcan, C.; Fântânariu, M.; Burtan, L.; Hriţcu, L.; Chiruţă, C.; Fernández-Mérida, L.; Rodríguez-Raposo, R.; Santana, J.J.; et al. New Ti-6Al-2Nb-2Ta-1Mo Alloy as Implant Biomaterial: In Vitro Corrosion and In Vivo Osseointegration Evaluations. *Mater. Chem. Phys.* **2020**, *240*, 122229. [CrossRef]

64. Trincă, L.C.; Mareci, D.; Souto, R.M.; Lozano-Gorrín, A.D.; Izquierdo, J.; Burtan, L.; Motrescu, I.; Vulpe, V.; Pavel, G.; Strungaru, S.; et al. Osseointegration Evaluation of ZrTi Alloys with Hydroxyapatite-Zirconia-Silver Layer in Pig's Tibiae. *Appl. Surf. Sci.* **2019**, *487*, 127–137. [CrossRef]
65. Ferracane, J.L. Resin Composite—State of the Art. *Dent. Mater.* **2011**, *27*, 29–38. [CrossRef] [PubMed]
66. Drummond, J.L. Degradation, Fatigue and Failure of Resin Dental Composite Materials. *J. Dent. Res.* **2008**, *87*, 710–719. [CrossRef] [PubMed]
67. Tzioupis, C.; Giannoudis, P.V. Prevalence of Long-Bone Non-Unions. *Injury* **2007**, *38*, S3–S9. [CrossRef]
68. Narayan, R. *Monitoring and Evaluation of Biomaterials and Their Performance In Vivo*; Woodhead Publishing: Sawston, UK, 2016; ISBN 978-0-08-100603-0.
69. Gao, H.; Huang, J.; Wei, Q.; He, C. Advances in Animal Models for Studying Bone Fracture Healing. *Bioengineering* **2023**, *10*, 201. [CrossRef] [PubMed]
70. Bozzini, C.; Picasso, E.O.; Champin, G.M.; Alippi, R.M.; Bozzini, C.E. Biomechanical Properties of the Mid-Shaft Femur in Middle-Aged Hypophysectomized Rats as Assessed by Bending Test. *Endocrine* **2012**, *42*, 411–418. [CrossRef]
71. Suckow, M.A.; Weisbroth, S.H.; Franklin, C.L. *The Laboratory Rat*; Elsevier: Amsterdam, The Netherlands, 2005; ISBN 978-0-12-074903-4.

life

MDPI

Article

Exploring Lacrimal Gland Tear Production in Sheep under General Anesthesia: Examining the Potential Impact of Utilizing 1% Hyaluronic Acid Ophthalmic Gel

Ruxandra Pavel *, Ioana Ene and Ruxandra Costea

Faculty of Veterinary Medicine, University of Agronomic Sciences and Veterinary Medicine of Bucharest, 011464 Bucharest, Romania; ioana.ene@fmvb.usamv.ro (I.E.)
* Correspondence: ruxandra.pavel@fmvb.usamv.ro

Abstract: The aim of the study is to assess the lacrimal gland tear production, using the Schirmer Tear Test (STT), in healthy sheep under general anesthesia and to explore the effects of applying 1% hyaluronic acid ophthalmic gel during general anesthesia. While STT values during anesthesia have been well documented in small animals such as cats and dogs, there seems to be a lack of information available for ruminants like sheep. This gap in the literature highlights the need for further research and exploration into tear production in sheep under anesthesia. The experimental research conducted on twelve adults female sheep provided valuable insights into tear production under anesthesia. By assessing tear production at various key time points the study was able to capture the changes in tear production throughout anesthesia and revealed a significant decrease in Schirmer Tear Test values in all sheep, following general anesthesia. The results showed that closing and taping the eye yielded determined better outcomes compared to administering a 1% lubricant ophthalmic gel. This finding suggests that eye care during anesthesia can impact tear production in sheep.

Keywords: sheep; anesthesia; lacrimal gland; tear production; ophthalmic gel

1. Introduction

Tears play a vital role in preserving eye health by eliminating foreign particles, supplying crucial nutrients to the avascular cornea, and containing immunoglobulins, lysozymes, and other essential proteins that help defend the eye [1].

Histologically, the lacrimal glands consist of acini, which comprise layers of myoepithelial cells with lightly stained secretory granules, along with prominent serous cells. These structures are encased in a rare and vascular stroma [2].

When discussing eye examinations, it is crucial to follow specific steps to ensure there is no compromise and no further complications arise [3]. The Schirmer Tear Test (STT) is a crucial examination in ophthalmology for assessing tear production and diagnosing various eye pathologies like keratitis or keratoconjunctivitis sicca [4]. It plays a significant role in determining the quality and quantity of tears, providing valuable insights into the overall ocular health of an individual.

Insufficient tear production leads to keratoconjunctivitis sicca (KCS), commonly referred to as "dry eye" or xerophthalmia, causing gradual inflammation of the cornea and conjunctiva [1,5,6]. Ocular discharge, because of this inflammation, is a common symptom observed in individuals with KCS, although the exact extent of its increase may vary and remain uncertain. Hence, understanding the baseline tear production in healthy sheep is crucial, and this can be performed by the STT, currently considered the most accurate method to measure tear production in animals [7].

Tear film is composed of three layers which are produced by the meibomian glands and the glands of Zeis that form the outer thin and fatty layers that prevent the underlying aqueous layer from evaporating or overflowing [8]. The middle aqueous layer, which constitutes over 60% of the entire tear film, is primarily responsible for its function and is the thickest among the three layers. The lacrimal gland, accessory lacrimal glands, and the nictitans gland are the ones that produce the aqueous layer [9]. The main function of the aqueous layer of the tear film is represented by preserving nutrition and oxygen supply to the cornea along with the protection of its surface from foreign bodies, epithelial waste, or other toxic substances [10]. Conjunctival goblet cells produce the inner mucous layer that has its main characteristic of transforming the hydrophobic epithelial surface into a hydrophilic one [11].

Corneal abrasions are one of the most common ophthalmological complications related to general anesthesia during surgeries that do not involve the eye. It is important to be aware of this risk and take necessary precautions to prevent it [12]. Other complications may include lagophthalmos, resulting in exposure keratopathy, diminished eyelid reflex, lowered basal tear production, and compromised stability of the corneal tear film [13].

Anesthesia is frequently performed in sheep and pigs for experimental research, employing a diverse approach to ensure effective sedation [14,15].

The effect of anesthetics on tear production is evaluated in different species, but in sheep, the current veterinary literature has few studies regarding normal values of tear production and how general anesthesia affects the STT measurements [16].

Reduced tear production may result from the interaction of general anesthetics with the parasympathetic nervous system. Alternatively, it could be due to diminished blood flow in the lacrimal gland or alterations in the tear-producing cells [17].

In our study, we aimed to compare tear production in sheep during anesthesia using two different protection methods: taping the eye vs. applying a 1% hyaluronic acid ocular gel. While we did not come across any specific research on eyelid closure in sheep during anesthesia, there are several studies in human medicine that delve into the causes of corneal injuries linked to anesthesia and assess the advantages and disadvantages of various corneal protection methods [18–23].

The primary hypothesis posits that general anesthesia leads to a significant decrease in tear secretion. The secondary hypothesis suggests that closing the eyelid during anesthesia will help prevent rapid evaporation of the tear film and facilitate a quicker return to normal values of STT compared to situations where 1% lubricant ophthalmic gel was used without other protection.

2. Materials and Methods

This study was in accordance with the Ethics Committee of the Faculty of Veterinary Medicine Bucharest and the methodology used adhered to the guidelines outlined in the Public Health Service Policy on the Humane Care and Use of Laboratory Animals (2015). The study was conducted on 12 Tsurcana female sheep (24 eyes) during the spring season, with a mean age and weight of 6.33 years old and 48.25 kg, respectively. All animals were housed under the same environmental, nutritional, and welfare conditions. A thorough and calm preanesthetic physical examination was performed, evaluating cardiovascular and respiratory functions, temperature, and hydration status. The color of the mucous membranes was assessed, and additionally blood samples were obtained to assess health status of the sheep, including a complete ophthalmologically exam. All sheep included in this study were clinically healthy.

Patient preparation included a 12 h fasting and a 6 h water restriction. Regurgitation is more likely if this preoperative fast period is not followed. Fasting before anesthesia helps minimize gas production from fermentation and prevents rumen bloating. This, in turn, reduces intraabdominal pressure on the diaphragm and major blood vessels, thereby preventing compromised cardiovascular function [14].

All sheep were premedicated with 0.2 mg/kg of midazolam (Midazolam SUN 1 mg/mL, Sun Pharmaceutical Industries Europe B.V., Cluj-Napoca, Romania), 5 mg/kg of ketamine (Ketamidor® 100 mg/mL, VetViva Richter GmbH, Wels, Austria), and 0.1 mg/kg of butorphanol (Butomidor® 10 mg/mL, VetViva Richter GmbH, Wels, Austria) intramuscularly

(IM). After premedication, a 20 G catheter was placed in the cephalic vein (Figure 1). The induction was performed with 3–6 mg/kg of propofol (Fresofol® 1% MCT/LCT, Fresenius Kabi Australia, Mount Kuring-gai, Australia) intravenously (IV).

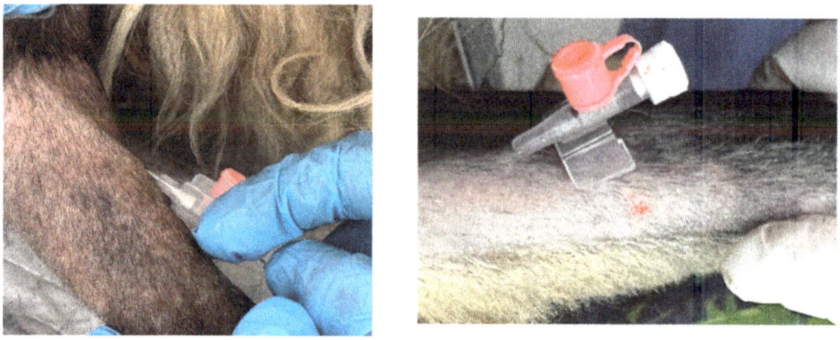

Figure 1. Sheep with a 20 G catheter in the cephalic vein.

Sheep were intubated and maintenance of anesthesia was performed with isoflurane (Isothesia, 1000 mg/g, Piramal Critical Care B.V., Voorschoten, The Netherlands) and oxygen 100% (Figure 2a,b).

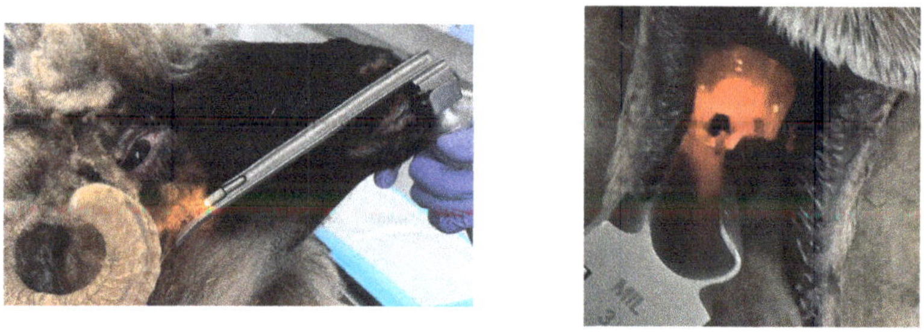

(**a**) (**b**)

Figure 2. (**a**) Intubation using a laryngoscope with a Miller blade; (**b**) visualization of the larynx.

Schirmer Tear Test (Schirmer-Tränentest, Vet Eickemeyer®, Tuttlingen, Germany) was performed for both eyes, with baseline values being recorded and expressed in millimeters per minute. During STT, the eyelids were gently closed, and the test strip was placed into the temporal third of the lower eyelid for a duration of 1 min (Figure 3a). Following the removal of the test strip, the length of the moistened region was promptly measured in millimeters (Figure 3b).

Tear production was measured at baseline T_0 (15 min before premedication). After 15 min of premedication, we recorded a second value of the STT (T_1).

For the third value of the STT, immediately after intubation of the sheep, we chose to tap the eyelids of the right eye (OD) with adhesive band, and for the left eye (OS), we instilated 2 drops of a lubricant ophthalmic gel with 1% hyaluronic acid and waited 15 min before measuring the tear secretion (T_2).

At the end of the surgery, we waited 15 min after extubation and we recorded a fourth STT value (T_3). The last measurement was performed for all sheep 24 h after the end of the surgery (T_4).

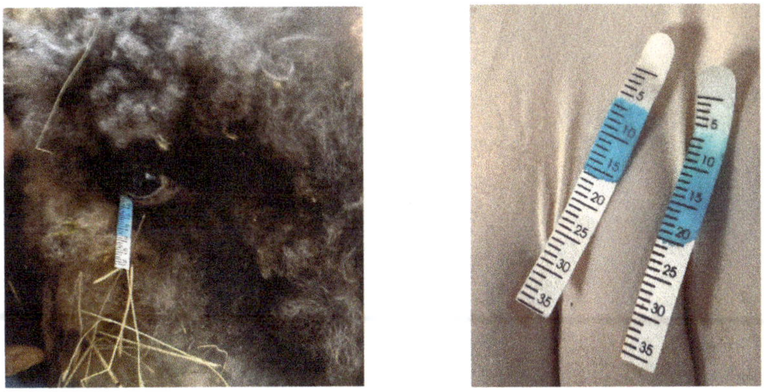

(**a**) (**b**)

Figure 3. (**a**) Placing the Schirmer Tear Test; (**b**) values of the Schirmer Tear Test.

The surgery of the sheep was represented by removing an electrode and electronic modules implanted subcutaneously in the right posterior leg, in the croup region, in a previous surgery. For this study, all sheep were positioned in sternal recumbency with the head straight (Figure 4).

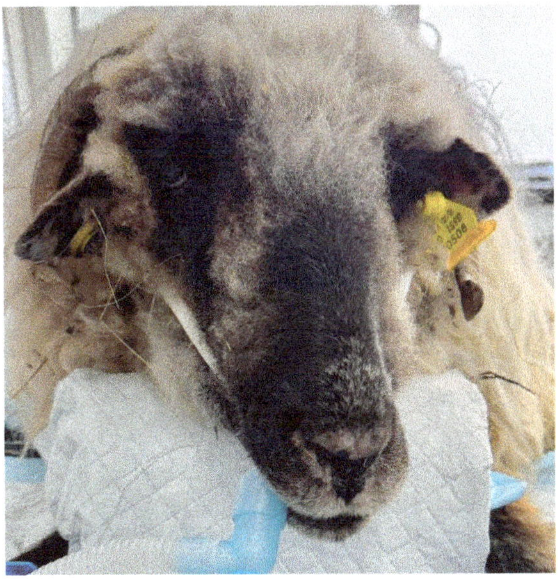

Figure 4. Head position during surgery, immediately after intubation.

All sheep had the same period of being under anesthesia, respectively, 1 h.

A statistical analysis of the data was performed using the software program DATAtab (DATAtab Team 2024. DATAtab: Online Statistics Calculator. DATAtab e.U. Graz, Austria. URL https://datatab.net, accessed on 12 July 2024) and Microsoft Excel (Version 16.82, 2024).

Mean values, along with the standard deviation (SD), were calculated for the STT results. All measurements were conducted around the same time of the day, between 9 am and 3 pm, to account for potential diurnal variations in tear production [24,25]. Additionally, the measurements were performed by the same individual.

A paired sample *t*-test was used to compare the STT values obtained before and after premedication, after taping the right eye (OD) and applying 2 drops of the 1% hyaluronic acid lubricant ophthalmic gel in the left eye (OS), after extubation, and after 24 h. We analyzed whether taping the eye without applying the ocular lubricants can influence the evaporation of the tear film and the return of the normal values faster. Statistical significance was defined as a p-value < 0.05.

3. Results

All measurements were completed successfully. It is known that during summer and spring, sheep tend to have higher STT values compared to winter, as described by Dedousi [26]. Previous research in horses and ponies has shown a tendency for increased tear scores in the STT during winter compared to those observed in summer, although these differences did not reach statistical significance [27]. Moreover, a previous study carried out on horses stated that the amount of tear secretion can vary according to the seasons [27]. All data were obtained in the spring season, within the same period of the day, from 9 am to 3 pm; therefore, changes in the daily cycle and season, which might influence the test results, most likely did not affect our STT results since the differences between STT values were observed in different seasons readings, early in the morning or late evening [4,24,28,29].

The mean values ± SD of basal tear production, expressed in millimeters/min, measured before premedication (T_0) on right (OD) and left eye (OS), were 19.08 ± 3.96 mm/min and 17.33 ± 2.53 mm/min, respectively (Figure 5a). The mean values ± SD in both eyes after 15 min of premedication (T_1) were 13.08 ± 3.31 mm/min for the right eye (OD) and 12 ± 3.43 mm/min for the left eye (OS) (Figure 5b).

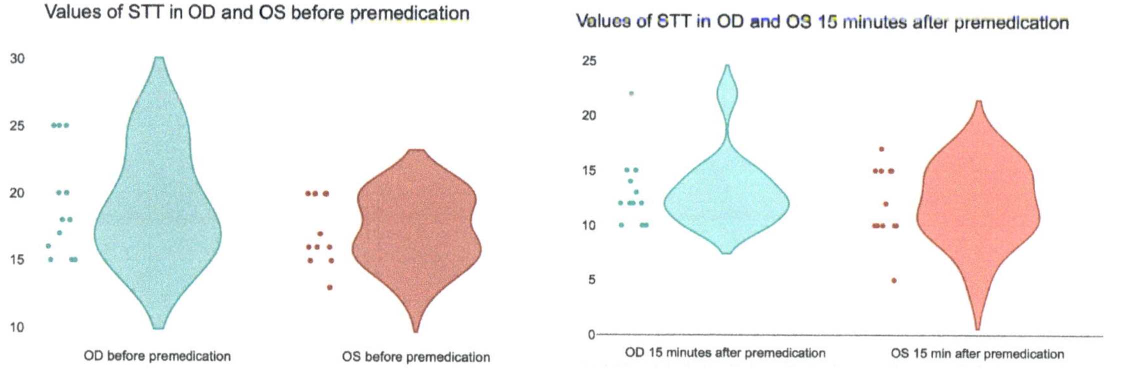

(**a**) (**b**)

Figure 5. (**a**) Values of the Schirmer Tear Test in the right (OD) and left eye (OS) before premedication; (**b**) values of the Schirmer Tear Test in the right and left eye 15 min after premedication.

The results of the *paired-t* test indicated that there is a significant large difference between the right eye (OD) before premedication (T_0) (M = 19.08, SD = 3.96) and after 15 min of premedication (T_1) (M = 13.03, SD = 3.31), $t(11) = 8.8$, $p < 0.001$ (Figure 6a). Regarding the left eye (OS), the results of the paired-*t* test indicates that there is a significant large difference between the STT values before premedication (T_0) (M = 17.33, SD = 2.53) and 15 min after premedication (T_1) (M = 12, SD = 3.43), $t(11) = 7.7$, $p < 0.001$ (Figure 6b).

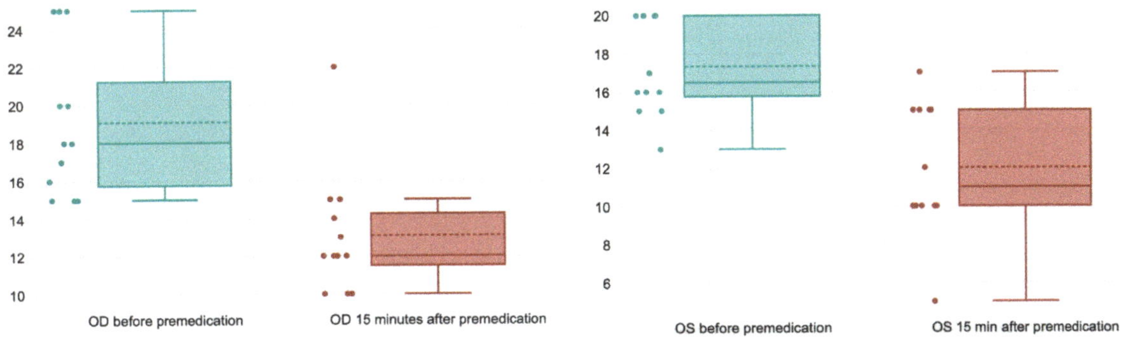

(**a**) (**b**)

Figure 6. (**a**) Values of the Schirmer Tear Test in the right eye (OD) before premedication and 15 min after premedication; (**b**) values of the Schirmer Tear Test in the left eye (OS) before premedication and 15 min after premedication.

The mean values ± SD 15 min after intubation and taping the right eye (OD) and after instilating two drops of ocular gel with 1% hyaluronic acid in the left eye (OS) (T_2) were 8.75 ± 2.70 and 3.58 ± 2.71 mm/min, respectively. The *t*-value is 4.67584, and the *p*-value is 0.000058. The result is *significant* at $p < 0.05$.

There was a *significant* decrease in tear secretion value in sheep who received only the ocular lubricant compared with the taped eye (Figure 7). Our hypothesis from our study is that closing the eyelid in the right eye (OD) during anesthesia prevents a rapid evaporation of the tear film, and a better outcome which was also seen in the values that we obtained.

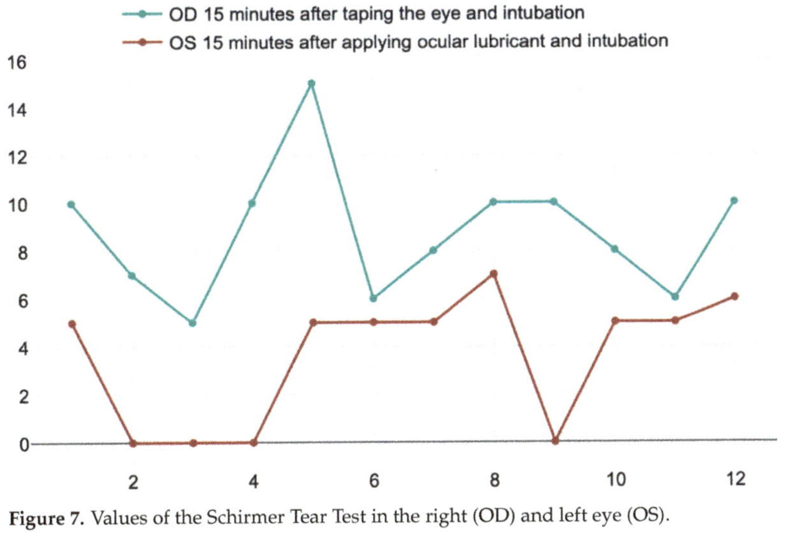

Figure 7. Values of the Schirmer Tear Test in the right (OD) and left eye (OS).

The mean values ± SD of the right (OD) and left eye (OS) after 15 min of extubation and returning of the palpebral reflexes were, respectively, 16 ± 3.71 mm/min and 10.25 ± 3.67 mm/min. There was a *significant* difference regarding the return of tear production (T_3) in the right eye (OD), which was taped, compared with the left eye (OS), which received only two drops of ocular gel with 1% hyaluronic acid. The *t*-value is 3.81253, and the *p*-value is 0.000476. The result is *significant* at $p < 0.05$ (Figure 8).

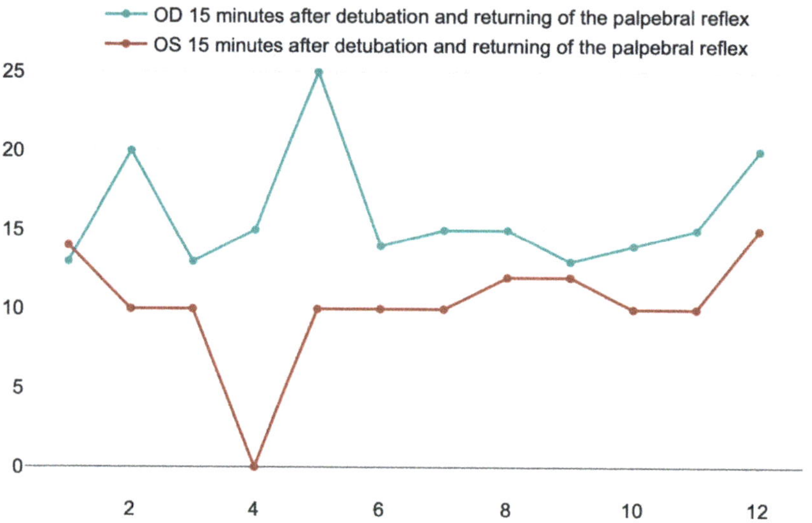

Figure 8. Values of the Schirmer Tear Test in the right (OD) and left eye (OS) 15 min after detubation and returning of the palpebral reflex.

After 24 h (T_4), we evaluated the return of the tear production values at baseline. In the right eye (OD), which was taped during anesthesia, the results of the paired-*t* test indicate that there is a *non-significant* small difference between the values of the STT before premedication ($M = 19.08$, $SD = 3.96$) and 24 h after premedication ($M = 18.03$, $SD = 2.42$), $t(11) = 1.1$, $p = 0.137$, compared with the values for the left eye (OS), who received the ocular lubricant where the results of the paired-*t* test indicated that there is a significantly large difference between the values of the STT before ($M = 17.33$, $SD = 2.53$) and after 24 h from premedication ($M = 14.16$, $SD = 1.89$), $t(11) = 4.4$, $p < 0.001$ (Figure 9).

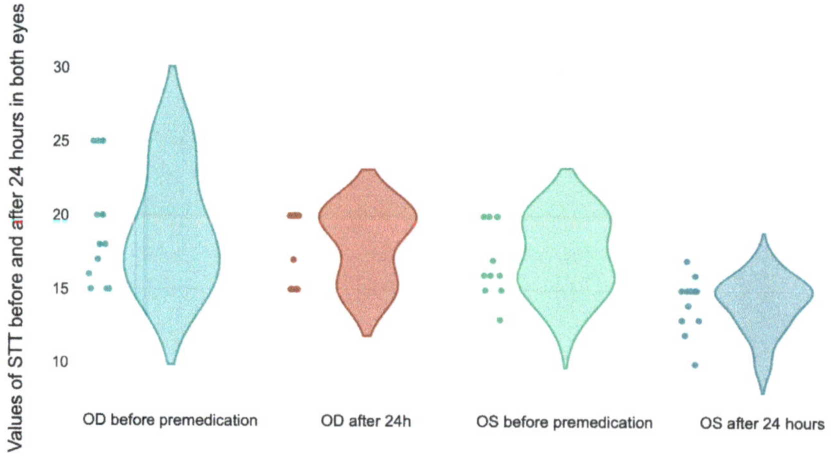

Figure 9. Values of the Schirmer Tear Test in both eyes before premedication and 24 h after premedication.

At the end of the study, each sheep had a Fluoresceine test (AIESI® Fluorescein Sodium Fluorescein Sterile Ophthalmic Strips for Eye Tone Test HOSPIFLUO STRIPS, Manufactured by CONTACARE OPHTHALMICS & DIAGNOSTICS, Gujarat, India) performed, achieving a negative result which established that none of them had any corneal abrasions and none were displaying other clinical symptoms like blepharospasm, ocular pain, and epiphora that could indicate an affected cornea.

All the mean values ± SD we obtained are displayed in the following table (Table 1).

Table 1. Values of mean and standard deviation of tear production in both eyes during different periods in time.

	STT Values in Both Eyes at Different Time Periods	Mean ± SD
T_0	OD before premedication OS before premedication	19.09 ± 3.96 mm/min 17.33 ± 2.53 mm/min
T_1	OD 15 min after premedication OS 15 min after premedication	13.08 ± 3.31 mm/min 12 ± 3.43 mm/min
T_2	OD 15 min after taping the eye and intubation OS 15 min after instilating the ocular lubricant and intubation	8.75 ± 2.70 mm/min 3.58 ± 2.71 mm/min
T_3	OD 15 min after detubation and returning of the palpebral reflex OS 15 min after detubation and returning of the palpebral reflex	16 ± 3.71 mm/min 10.25 ± 3.67 mm/min
T_4	OD after 24 h OS after 24 h	18.08 ± 2.42 mm/min 14.16 ± 1.89 mm/min

4. Discussion

Based on the findings of this study, the results suggest a decrease in tear production during anesthesia with midazolam, butorphanol, ketamine, and the other results suggest that closing the eyelid in the right eye (OD) during anesthesia prevents a rapid evaporation of the tear film and a fast return to normal values of STT compared with the left eye (OS), where we only used two drops of the ocular lubricant.

To our knowledge, the results of this study are not comparable within the literature, as there are no other studies that have evaluated the impact on tear production after the administration of this sedative agents nor the impact of taping the eye compared with instilating the ocular lubricant with 1% hyaluronic acid in the other eye, in Tsurcana sheep.

The mean baseline values of the STT (T_0) before premedication were, respectively, 19.08 ± 3.96 mm/min in the right eye (OD) and 17.33 ± 2.53 mm/min in the left eye (OS). Previous research has indicated that the average tear secretion (TS) rates in various sheep breeds are as follows: Sanjabi sheep have a baseline rate of 18.52 ± 2.55 mm/min, as reported by Ghaffari et al. [30]; Romanov sheep have a lower baseline rate, with 11.59 ± 4.07 mm/min [31]; and Merinos sheep have a higher rate of 18.80 ± 1.82 mm/min [32]. Variations in filter papers, the positioning of strips in the conjunctival sac, and the person conducting the test can influence TS values due to differences in absorptive capacities, as noted by Ghaffari et al. [30] and Rothschild et al. [33]. Moreover, a previous study carried out on horses stated that the amount of TS can vary according to the season [27]. All our data were obtained in the spring season, at the same period of day, from 9 am to 3 pm; therefore, changes in the daily cycle and season, which might influence the test results, most likely did not affect our STT results since the differences between STT values were observed in readings from different seasons, and early in the morning or in the late evening [4,24,28,29].

4.1. *Effects of the Anesthetic Drugs on Tear Production*

The precise mechanism behind the reduction in TS in sheep remains uncertain. However, a prior investigation revealed that the intramuscular administration of a combination of sedatives and opioid led to a decrease in TS levels in dogs [17]. In our study, the use of opioids, sedatives, and dissociative anesthetics led to a fast and constant decrease in the tear secretion following the premedication, induction, and maintenance phase.

Dodam et al. [17] proposed that the reduction in TS levels could be attributed primarily to one or a combination of the following mechanisms: the central effects of these drugs on the autonomic regulation of tear production, effective pain relief, vasoconstriction of the tear gland, and alterations in metabolism at the cellular level of the gland. Additionally, another potential explanation for the decrease in TS levels could be that the administration of sedatives resulted in a reduced blinking rate and increased evaporation of tears, as suggested by Leonardi et al. [34].

4.2. Effect of 1% Hyaluronic Acid Compared with the Eye Closed

Although sodium hyaluronate positively impacts tear film stability, a recent pilot study revealed that among two groups treated with hyaluronic acid (HA), the group receiving a lower concentration (0.25%) had higher STT readings compared to the group receiving a higher concentration (1%) [35]. In our study, closing the eyelid in the right eye (OD) during anesthesia prevented a rapid evaporation of the tear film and a fast return to normal values of STT compared with the left eye (OS), where we only use the ocular lubricant with 1% hyaluronic acid. While ocular lubricants do not increase tear production rates, their use is recommended to reduce corneal dehydration, enhance corneal wettability, and protect the cornea [36].

5. Conclusions

The data obtained in the present research provide important information, helping the anesthesiologists to better manage the effect of anesthetic drugs regarding tear production. Anesthesia is required for several surgery procedures, and a good understanding of how the drugs affect the STT and how the cornea can be protected is important for good perianesthetic management. This study suggested that the protocol used with midazolam, butorphanol, and ketamine decreases tear production shortly after intramuscular administration in sheep. There are numerous methods available to protect the cornea, but the one that we studied by taping the eyelid of the right eye (OD) and putting ocular lubricant with 1% hyaluronic acid in the left eye (OS) suggested better STT values and a faster return to baseline after the palpebral reflexes appeared in the right eye (OD) compared with the left one (OS).

Author Contributions: Conceptualization, R.P. and R.C.; methodology, R.P. and R.C.; software, R.P. and R.C.; validation, R.P., R.C. and I.E., investigation, R.P., R.C. and I.E.; resources, R.P. and R.C.; data curation, R.P. and R.C.; writing—original draft preparation, R.P. and R.C.; writing—review and editing, R.P., R.C. and I.E. All authors have read and agreed to the published version of the manuscript.

Funding: This research received no external funding.

Institutional Review Board Statement: This study complied with the ethical principles in scientific research regarding the use of animals in procedures, submitted and favourably approved by the Bioethics Committee No. 33/1.2.2024 of the Faculty of Veterinary Medicine of Bucharest, România.

Informed Consent Statement: Not applicable.

Data Availability Statement: The data generated in this study are presented in the tables of this article. For any further information, the reader can contact the authors.

Acknowledgments: To all the colleagues from the Faculty of Veterinary Medicine, University of Agronomic Sciences and Veterinary Medicine of Bucharest, who helped us in completing many cases, which were the basis of our experiences necessary in writing this article.

Conflicts of Interest: The authors declare no conflicts of interest.

References

1. Gelatt, K.N.; Gilger, B.C.; Kern, T.J. *Veterinary Ophthalmology*, 5th ed.; Wiley-Blackwell: Ames, IA, USA, 2013; pp. 568–573, 583–590.
2. Dalga, S.; Aksu, S.I.; Aslan, K.; Deprem, T.; Uğran, R. Anatomical and Histological Structures of Eye and Lacrimal Gland in Norduz and Morkaraman Sheep. *Turk. J. Vet. Anim. Sci.* **2022**, *46*, 336–346. [CrossRef]
3. Featherstone, H.J.; Heinrich, C.L. Ophthalmic Examination and Diagnostics. In *Veteinary Ophthalmology*, 5th ed.; Gelatt, K.N., Gilger, B.C., Kern, T.J., Eds., Two Volume Set; Wiley-Blackwell: Oxford, UK, 2013; pp. 533–702.

4. Ribeiro, A.; Piso, D.; Padua, I.; Silva, M.; Laus, J. Intraocular Pressure and Tear Secretion in Saanen Goats with Different Ages. *Pesqui. Veterinária Bras.* **2010**, *30*, 798–802. [CrossRef]
5. Van Kampen, K.R.; James, L.F. Ophthalmic lesions in locoweed poisoning of cattle, sheep, and horses. *Am. J. Vet. Res.* **1971**, *32*, 1293–1295.
6. Williams, D.L.; Tighe, A. Immunohistochemical Evaluation of Lymphocyte Populations in the Nictitans Glands of Normal Dogs and Dogs with Keratoconjunctivitis Sicca. *Open Vet. J.* **2018**, *8*, 47–52. [CrossRef]
7. Kurt, B.K.; Bulut, O.; Bozkan, Z.; Şen, Z.B.; Belge, A. Determination of Tear Volume and Intraocular Pressure in Saanen Goat and Sakiz Sheep in Similar Environmental Conditions. *Int. J. Vet. Anim. Res. (IJVAR)* **2021**, *4*, 78–82.
8. Samuelson, D.A. Ophthalmic anatomy. In *Veterinary Ophthalmology*; Gelatt, K.N., Ed.; Blackwell: Ames, IA, USA, 2007; pp. 37–148.
9. Gum, G.G.; Gelatt, K.N.; Esson, D.W. Physiology of the eye. In *Veterinary Ophthalmology*; Gelatt, K.N., Ed.; Blackwell: Ames, IA, USA, 2007; pp. 149–182.
10. Rolando, M.; Zierhut, M. The Ocular Surface and Tear Film and Their Dysfunction in Dry Eye Disease. *Surv. Ophthalmol.* **2001**, *45*, S203–S210. [CrossRef]
11. Ehlers, N. The Precorneal Film. Biomicroscopical, Histological and Chemical Investigations. *Acta Ophthalmol. Suppl.* **1965**, (Suppl. 81), 81–134.
12. Grixti, A.; Sadri, M.; Watts, M.T. Corneal Protection during General Anesthesia for Nonocular Surgery. *Ocul. Surf.* **2013**, *11*, 109–118. [CrossRef]
13. Kaye, A.D.; Renschler, J.S.; Cramer, K.D.; Anyama, B.O.; Anyama, E.C.; Gayle, J.A.; Armstead-Williams, C.M.; Mosieri, C.N.; Saus, J.A.; Cornett, E.M. Postoperative Management of Corneal Abrasions and Clinical Implications: A Comprehensive Review. *Curr. Pain Headache Rep.* **2019**, *23*, 48.
14. Costea, R.; Pavel, R.; Girdan, G.; Ene, I.; Posastiuc, F.; Micsa, C.; Constantin, T.; Togoe, D.; Diaconescu, A. Partial Intravenous Anesthesia with Isoflurane and Alfaxalone for an Adult Sheep Undergoing Soft Tissue Surgery. *Sci. Work. Ser. C Vet. Med.* **2023**, *69*, 57–60.
15. Costea, R.; Ene, I.; Pavel, R. Pig Sedation and Anesthesia for Medical Research. *Animals* **2023**, *13*, 3807. [CrossRef]
16. Peche, N.; Köstlin, R.; Reese, S.; Pieper, K. Postanaesthetic Tear Production and Ocular Irritation in Cats. *Tierärztliche Prax. Kleintiere* **2015**, *43*, 75–82. [CrossRef]
17. Dodam, B. Effects of Intramuscular Sedative and Opioid Combinations on Tear Production in Dogs. *Vet. Ophthalmol.* **1998**, *1*, 57–59. [CrossRef]
18. Smolle, M.; Keller, C.; Pinggera, G.; Deibl, M.; Rieder, J.; Lirk, P. Clear Hydro-Gel, Compared to Ointment, Provides Improved Eye Comfort after Brief Surgery. *Can. J. Anesth.* **2004**, *51*, 126. [CrossRef] [PubMed]
19. Siffring, P.; Poulton, T. Prevention of Ophthalmic Complications during General Anesthesia. *J. Am. Soc. Anesthesiol.* **1987**, *66*, 569–570. [CrossRef]
20. Schmidt, P.; Bøggild-Madsen, N.B. Protection of the Eyes with Ophthalmic Ointments during General Anaesthesia. *Acta Ophthalmol.* **1981**, *59*, 422–427. [CrossRef]
21. Orlin, S.E.; Kurata, F.K.; Krupin, T.; Schneider, M.; Glendrange, R.R. Ocular Lubricants and Corneal Injury during Anesthesia. *Anesth. Analg.* **1989**, *69*, 384–385.
22. Grover, V.K.; Kumar, K.V.; Sharma, S.; Sethi, N.; Grewal, S.P. Comparison of Methods of Eye Protection under General Anaesthesia. *Can. J. Anaesth.* **1998**, *45*, 575–577. [CrossRef]
23. Bøggild-Madsen, N.B.; Bundgarrd-Nielsen, P.; Hammer, U.; Jakobsen, B. Comparison of Eye Protection with Methylcellulose and Paraffin Ointments during General Anaesthesia. *Can. Anaesth. Soc. J.* **1981**, *28*, 575–578. [CrossRef]
24. Kovalcuka, L.; Malniece, A. Measurement of Tear Production and Intraocular Pressure in Clinically Conscious Normal Captive Red Deer (*Cervus elaphus*). *Animals* **2024**, *14*, 940. [CrossRef]
25. Orhan, A.; Bozkan, Z. Evaluation of Normal Tear Volume and Intraocular Pressure in Saanen Goats at Different Periods. *Int. J. Vet. Anim. Res. (IJVAR)* **2022**, *5*, 154–158.
26. Dedousi, A.; Karatzia, M.A.; Katsoulos, P.D. Reference Values of Schirmer Tear Test in Sheep and the Effect of Season on the Test Results. *Acta Vet. Hung.* **2019**, *67*, 553–560. [CrossRef]
27. Beech, J.; Zappala, R.; Smith, G.; Lindborg, S. Schirmer Tear Test Results in Normal Horses and Ponies: Effect of Age, Season, Environment, Sex, Time of Day and Placement of Strips. *Vet. Ophthalmol.* **2003**, *6*, 251–254. [CrossRef]
28. Del Sole, M.J.; Sande, P.H.; Bernades, J.M.; Aba, M.A.; Rosenstein, R.E. Circadian Rhythm of Intraocular Pressure in Cats. *Vet. Ophthalmol.* **2007**, *10*, 155–161. [CrossRef]
29. Giannetto, C.; Piccione, G.; Giudice, E. Daytime Profile of the Intraocular Pressure and Tear Production in Normal Dog. *Vet. Ophthalmol.* **2009**, *12*, 302–305. [CrossRef] [PubMed]
30. Ghaffari, M.S.; Shojaei, M.; Sabzevari, A.; Khorami, N. Reference Values for Intraocular Pressure and Schirmer Tear Test in Clinically Normal Sanjabi Sheep. *Small Rumin. Res.* **2011**, *97*, 101–103. [CrossRef]
31. Akgul, M.B. The effect of late pregnancy on tear secretıon ın romanov sheep. *Int. J. Inf. Res. Rev.* **2019**, *6*, 6078–6080.
32. Işler, C.; Altuğ, M.; Kilic, S. Evaluation of Tear Fluid Secretion and Intraocular Pressure in Normal Merinos Sheep and Saanen Goats. *Rev. Med. Vet.* **2013**, *164*, 278–282.
33. Rothschild, C.M.; Sellon, D.C.; Bryan, G.M.; Gay, J.M.; Hines, M.T. Effects of Trimethoprim–Sulfadiazine on Tear Production and the Fluctuations of Schirmer Tear Test Values in Horses. *Vet. Ophthalmol.* **2004**, *7*, 385–390. [CrossRef] [PubMed]

34. Leonardi, F.; Costa, G.L.; Stagnoli, A.; Zubin, E.; Boschi, P.; Sabbioni, A.; Simonazzi, B. The Effect of Intramuscular Dexmedetomidine-Butorphanol Combination on Tear Production in Dogs. *Can. Vet. J.* **2019**, *60*, 55. [PubMed]
35. Di Palma, C.; Micieli, F.; Lamagna, B.; Nieddu, A.; Uccello, V.; Fatone, G.; Vesce, G. Schirmer Tear Test Value and Corneal Lesions' Incidence during General Anesthesia for Non-Ophthalmic Surgery in Non-Brachycephalic Dogs: A Pilot Study Comparing Three Different Lubricant Eye Drop Formulations. *Vet. Sci.* **2020**, *7*, 25. [CrossRef] [PubMed]
36. Aragona, P.; Di Stefano, G.; Ferreri, F.; Spinella, R.; Stilo, A. Sodium Hyaluronate Eye Drops of Different Osmolarity for the Treatment of Dry Eye in Sjögren's Syndrome Patients. *Br. J. Ophthalmol.* **2002**, *86*, 879–884. [PubMed]

life

MDPI

Article

Detection of Ochratoxin A in Maize and Its Potential Impact on Avian Pathology in Romanian Farms

Silviu-Ionut Beia [1,†], Violeta Alexandra Ion [2,*,†], Elvira Gagniuc [3,4,*], Oana-Crina Bujor [2], Elena Ştefania Ivan [2], Andreea Barbu [2], Elena Pitoiu [4], Violeta Elena Beia [5] and Liliana Bădulescu [2]

1 Faculty of Management and Rural Development, University of Agronomic Sciences and Veterinary Medicine of Bucharest, 59 Marasti Blvd., 011464 Bucharest, Romania; beia.silviu@managusamv.ro
2 Research Center for Studies of Food Quality and Agricultural Products, University of Agronomic Sciences and Veterinary Medicine of Bucharest, 59 Marasti Blvd., 011464 Bucharest, Romania; oana.bujor@qlab.usamv.ro (O.-C.B.); elena.ivan@qlab.usamv.ro (E.Ş.I.); andreea.stan@qlab.usamv.ro (A.B.); liliana.badulescu@qlab.usamv.ro (L.B.)
3 Department of Pathology and Forensic Medicine, University of Agronomic Sciences and Veterinary Medicine of Bucharest, 59 Marasti Blvd., 011464 Bucharest, Romania
4 Synevovet Laboratory, 81 Pache Protopopescu Blvd., 021408 Bucharest, Romania; elena.pitoiu@synevo.com
5 National Sanitary Veterinary and Food Safety Authority, 1 Piaţa Presei Libere, 013701 Bucharest, Romania; beia.violeta@ansvsa.ro
* Correspondence: violeta.ion@qlab.usamv.ro (V.A.I.); elvira.gagniuc@fmvb.usamv.ro (E.G.)
† These authors contributed equally to this work.

Abstract: Ochratoxin A (OTA) is a nephrotoxic mycotoxin that commonly contaminates maize, posing significant health risks to both poultry and humans. In this study, a rapid and sensitive method utilizing ultra-performance liquid chromatography coupled with fluorescence detection (UPLC-FLD) was developed for the quantification of OTA levels in maize. The method utilizes immunoaffinity column purification for improved specificity. Accuracy and precision were validated in line with European Union Reference Laboratory (EURL-MP) guidelines, meeting regulatory standards for linearity, trueness, detection and quantification limits, precision, and uncertainty, as per European Commission Regulation (EC) No. 401/2006 and its amendments. The method demonstrated an average recovery rate of 116.78% for maize, with RSD_{wR} values (within-laboratory reproducibility) of 12.72%. Furthermore, OTA occurrence and its possible effects were investigated in several farms in South Romania, where necropsy and histopathological analyses of poultry revealed severe kidney damage, including renal tubular degeneration.

Keywords: OTA; mycotoxins; method validation; maize; histopathology; ultra-performance liquid chromatography; immunoaffinity column

1. Introduction

Mycotoxins are natural contaminants produced as secondary metabolites by certain genera of filamentous fungi, primarily *Aspergillus*, *Penicillium* [1], and *Fusarium*. More than 200 fungal species and approximately 300 distinct fungal metabolites have been identified as mycotoxins, including aflatoxins (AFs), ochratoxin A (OTA), zearalenone (ZEA), and fumonisins [2]. These fungi are generally classified into "field fungi", like *Fusarium*, which infect crops during growth, and "storage fungi", such as *Aspergillus* and *Penicillium*, which contaminate grains during storage [3,4]. *Aspergillus* is a large genus with over 100 species, some of which thrive in soil or decaying vegetation, while others are responsible for plant diseases and storage contamination. Mycotoxins pose serious health risks to humans and animals due to their toxic effects [5]. OTA is classified as a possible human carcinogen (Group 2B) and has been linked to a range of toxic effects, including hepatotoxicity, teratogenicity, neurotoxicity, immunotoxicity, and enterotoxicity in various animal species [6]. The European Union (EU) has classified mycotoxins as priority substances, recognizing them as more hazardous to health than common food and feed contaminants like pesticides, preservatives, and food additives [7,8]. Common sources of exposure include cereals, nuts, dried fruits, spices, coffee [9–11], grapes, fish [12], wine, beer [13], mollusks, herbs, and feed materials [14]. Given the significant impact of mycotoxins, particularly OTA, on agricultural products, it is crucial to explore these toxins in greater detail due to their serious implications for food safety and animal health. The fungi that produce OTA can thrive in temperatures ranging from 0 to 37 °C [13], and OTA is highly thermally stable, making its removal from feed challenging even at temperatures up to 250 °C [15]. Storage conditions play a critical role in preventing OTA contamination in agricultural products. High humidity and extended storage periods encourage fungal growth and OTA production, while low humidity and adequate ventilation reduce these risks. Effective storage practices include using airtight, moisture-proof containers, maintaining clean and mold-free environments, and thoroughly drying crops before storage. However, many small local farmers remain unaware of these practices, and without proper aeration their bulk storage methods often result in fungal contamination by spring.

Numerous analytical methods have been developed to investigate OTA, including capillary electrophoresis (CE) [16], gas chromatography-mass spectrometry (GC-MS) [17,18], and enzyme-linked immunosorbent assay (ELISA) [19]. However, most of the proposed methods rely on liquid chromatography (LC) coupled with various detection systems, such as fluorescence detection (LC-FLD) [20], tandem mass spectrometry (LC-MS/MS) [21], and high-resolution mass spectrometry (LC-HRMS) [22].

Sample extraction for OTA typically involves liquid–liquid extraction (LLE) or solid–liquid extraction (SLE), followed by cleanup with immunoaffinity columns (IAC) or solid-phase extraction (SPE) cartridges [22]. For OTA determination, traditional methods involve mixing a ground sample with an acetonitrile–water mixture, filtering, and applying the extract to a purification column [17]. SPE effectively isolates mycotoxins using sorbents such as C8, C18, aluminum oxide, and silica gel. Advanced SPE techniques, including magnetic SPE (m-SPE) and dispersive SPE (d-SPE), have been utilized to detect numerous mycotoxins and pesticides [21,23]. Additionally, the QuEChERS (Quick, Easy, Cheap, Effective, Rugged, and Safe) method simplifies mycotoxin detection by combining extraction, separation, and purification in one run, offering high throughput and significant recovery [21]. Understanding the extent and implications of OTA contamination relies heavily on the effectiveness of the analytical methods used to detect and measure this mycotoxin. The accurate detection of OTA in food and feed is crucial due to its significant health risks. Despite advances in analytical techniques, the primary concern remains the potential impact of OTA on animal and human health.

Feed ingestion is the primary route of mycotoxin exposure in animals [24]. Various mycotoxins, including aflatoxin B1 (AFB1), aflatoxin M1 (AFM1), citrinin (CIT), dihydrocitrinone (DH-CIT), deoxynivalenol (DON) conjugates, enniatin B (ENNB), and OTA, have been detected in animal biological fluids [25,26]. This presence suggests that animals and humans may also be exposed to mycotoxins through skin contact or inhalation of aerosols [27].

Poultry exhibit a lower absorption rate of OTA but are more sensitive to its effects compared to mammals [5]. OTA exerts significant nephrotoxic effects on broilers, leading to kidney enlargement and hyperplasia of the tubular epithelium [28,29]. It also adversely affects the immune system of birds, causing leukocytopenia [5,30]. The effect of OTA on growth includes a decreased body weight, poor feed conversion ratio, stunted growth, low-quality eggshells, reduced egg production, and immunosuppression. Furthermore, it can cause liver damage and affects gastrointestinal health and microbial communities [30, 31]. Clinical signs of OTA toxicity in poultry include weakness, anemia, decreased feed consumption, reduced growth rates, poor feathering, and elevated mortality rates at high dietary concentrations [32,33]. Pathophysiological changes associated with OTA exposure include decreased urine concentration, a reduced glomerular filtration rate, impaired proximal tubular function, and renal degeneration with ultrastructural alterations

Additionally, studies have reported increases in the relative weights of the liver, spleen, pancreas, proventriculus, gizzard, and testes in poultry fed OTA [34].

The aim of this study was to develop and implement a rapid and reliable method for the qualitative and quantitative detection of mycotoxins in maize used in intensive farming systems. Access to rapid diagnostic methods enables farmers to obtain timely toxicological results for maize, especially in cases where mycotoxicosis is suspected, allowing for swift intervention to minimize economic losses. Currently, in Romania, mycotoxicological analysis of maize requires a lengthy processing time, with an average confirmation period of approximately 21 days. The method developed in this study significantly reduces this timeframe to just 2 days, offering a faster and more efficient approach to OTA detection. The UPLC-FLD method was validated in line with European Union Reference Laboratory (EURL-MP) [35] guidelines, ensuring accuracy and regulatory compliance. This study hypothesized a correlation between OTA levels in maize and poultry health issues, such as increased mortality and specific pathological changes, as observed in Romanian farms where chickens consumed mold-contaminated maize. Histopathological examinations of tissue samples from recently deceased broiler chickens revealed potential OTA-related organ damage, particularly in the kidneys, along with reduced lymphocyte populations in lymphoid organs, suggesting compromised immune responses due to chronic OTA exposure.

2. Materials and Methods

2.1. Chemicals and Reagents

The standard solution was prepared using an OTA reference material of 100 µg/mL concentration in methanol purchased from LGC Limited Standards (Middlesex, TW11 0LY, London, UK). Methanol and acetonitrile (HPLC grade) were purchased from Honeywell (Riedel-de Haën, Seelze, Germany); glacial acetic acid was purchased from Fisher Scientific (Pittsburgh, PA, USA).

Potassium chloride was purchased from Scharlab S.L. (Sentmenat, Barcelona, Spain); sodium chloride, potassium dihydrogen phosphate, and disodium hydrogen phosphate were purchased from Carl Roth GmBH (Karlsruhe, Germany). OTAClean for Ochratoxin A immunoaffinity columns were obtained from LCTech GmbH (Obertaufkirchen, Germany). Phosphate-buffered saline (PBS, pH 7.2) was prepared by adding potassium chloride (0.20 g), potassium dihydrogen phosphate (0.20 g), anhydrous disodium hydrogen phosphate (1.16 g), and sodium chloride (8.00 g) to 900 mL of water. The pH was adjusted with 0.1 M NaOH and 0.1 M HCl. The volume was made up to 1 L with water. Tissue processing reagents included buffered 10% formaldehyde (Laurypath for Epredia, 69630 Chaponost, France), absolute ethyl alcohol, 70% and 95% ethyl alcohol, xylene (Laurypath for Epredia, 69630 Chaponost, France), paraffin wax (Histoplast LP, Epredia, Portsmouth, NH, USA), and hematoxylin and eosin (ST Infinity HE Staining kit, Leica Biosystems, Richmond, IL, USA).

2.2. Sample Collection

Several farms in Nuci and Moara Domnească (Ilfov County), Coșereni and Fierbinți (Ialomița County), and Lunca (Buzău County) were investigated for OTA contamination in maize used for poultry feeding (Table 1). These investigations were prompted by farmers' reports of unusually high poultry mortality rates, leading them to submit maize samples for OTA analysis. Samples from maize stored in bulk were collected in accordance with Commission Regulation (EU) No. 2782/2023, which amends the requirements of Regulation (EC) No. 401/2006 [36], to ensure that each sample represented its respective batch.

A total of forty incremental samples were collected from various points (upper, middle, and lower areas) to capture the entire mass of stored maize. These incremental samples were then combined to form an aggregate sample of approximately 4 kg. From the aggregate sample, two subsamples of at least 0.5 kg each were taken for examination and transported under optimal conditions to the research center's laboratory for mycotoxin analysis. Next, the samples were milled and sieved (0.5 mm) to produce representative test samples.

Table 1. Samples used for OTA analysis.

Sample Code	Variety	Storage Conditions	Location
M1	Carioca	Bulk in storage	Moara Domnească/ Ilfov
M2	P9889	Classic horizontal hall	Nuci/Ilfov
M3	P9889	Classic horizontal hall	Nuci/Ilfov
M4	P9889	Classic horizontal hall	Nuci/Ilfov
M5	P9889	Vertical silo made of galvanized sheet metal	Lunca/Buzău
M6	P9889	Horizontal silo made of concrete and masonry, warehouse type	Lunca/Buzău
M7	P9889	Bulk in storage	Coşereni/Ialomiţa
M8	P9889	Bulk in storage	Fierbinţi/Ialomiţa
M9	P9889	Bulk in storage	Fierbinţi/Ialomiţa

Among the investigated farms, the one with the highest OTA contamination, which was also selected for histopathological examination, was located in Nuci, Ilfov County. This farm housed 27,000 Ross 308 broiler chickens in a 1000 m^2 facility. According to the farmer, the feed was switched during the second week to a maize-based diet specifically formulated for the chickens' growth phase, which contained 60% maize.

The chickens were fed ad libitum, with continuous access to food and water. By the third week, the farm's mortality rate had reached 3%. However, shortly after the introduction of the new feed, mortality rates spiked, rising to 10–15% by the fourth week. The chickens also exhibited a marked decline in weight gain and deterioration in overall health. Clinical signs such as lethargy, anorexia, diarrhea, refusal to feed, dehydration, and damaged plumage became increasingly apparent. To determine the cause of the increased mortality, necropsies were conducted on 10 recently deceased broiler chickens. Samples from the kidneys, liver, thymus, bursa of Fabricius, and spleen were collected from 32-day-old Ross 308 chickens for histopathological examination.

2.3. OTA Extraction and Clean-Up

The extraction and cleanup of OTA in maize were performed using a modified procedure based on the OTAClean immunoaffinity column from LCTech GmbH, Obertaufkirchen, Germany [37]. In brief, 2 g of NaCl and 100 mL of an 80:20 (v/v) methanol solution were added to 20 g of a ground maize sample. The mixture was stirred for 1 h and then filtered. A 12 mL portion of the filtrate was diluted with 48 mL of PBS (pH 7.4) before passing through the immunoaffinity column, which contains immobilized monoclonal antibodies specific to OTA and offers a recovery rate exceeding 90% for the OTA standard, as stated in the certificate of analysis. After the extract fully passed through the column, it was washed with 10 mL of water, and the OTA was eluted using 1.5 mL of methanol. The collected eluate was refrigerated for subsequent analysis.

2.4. UPLC-FLD Instrument and Chromatographic Analysis

A Waters Acquity I chromatographic system (Milford, MA, USA) was used for OTA analysis, equipped with a UPLC binary pump, autosampler, injection system, and fluorescence detector. The analysis was performed under isocratic conditions, with a total runtime of 3 min. For OTA separation, a Zorbax Eclipse Plus C18 column (100 mm × 2.1 mm, 1.8 μm) (Agilent Technologies, Santa Clara, CA, USA) was employed. The mobile phase consisted of acetonitrile/water/acetic acid (70:30:1, $v/v/v$), with a flow rate of 0.5 mL/min and an injection volume of 5 μL. Detection was carried out at an excitation wavelength of 330 nm and an emission wavelength of 460 nm. During the analysis, the samples were maintained at 20 °C, and the column temperature was set to 30 °C.

2.5. OTA Method Validation

The method validation process evaluated several critical parameters, including the linearity, limit of quantification (LOQ), repeatability relative standard deviation (RSD_r), within-laboratory reproducibility relative standard deviation (RSD_{WR}), and recovery (R), in compliance with Commission Regulation (EU) No. 2782/2023, which amends the requirements of Regulation (EC) No. 401/2006 [36]. Performance criteria were also benchmarked against the proposed guidelines from the draft version of the EURL-MP [35].

To assess linearity, seven concentrations of OTA standard solutions (ranging from 0.5 to 10 ng/mL in methanol: acetic acid, 98:2, *v*/*v*) were used. Regression analysis was conducted to explore the relationship between OTA concentration (independent variable) and fluorescence detector intensity (dependent variable). Data analysis was performed using Microsoft Excel (version 2013), utilizing the Analysis ToolPak's Regression tool. Key statistical outputs included the correlation coefficient (Multiple R), the coefficient of determination (R^2), and the adjusted R^2. The ANOVA table was used to test the overall significance of the model, while the coefficient table provided detailed insights into variable relationships. Residual analysis was performed to confirm model fit and validate regression assumptions, with all statistical tests conducted at a significance level of $\alpha = 0.05$.

According to the EURL-MP guidelines [35], the limit of quantification (LOQ) was established at levels lower than 0.5 times the maximum level (ML). Method validation was conducted on maize flour quality control material (QC) sourced from Fera Science Ltd. (Sand Hutton, York, UK), which had an assigned ochratoxin A (OTA) value of 0.810 μg/kg. This assigned value served as the LOQ for the analysis, indicating that concentrations below this threshold may not be reliably quantified. The acceptable $|z|$-score range for this QC material was 0.454–1.166 μg/kg. Quality control procedures for the maize samples were performed through extraction and cleanup in accordance with the developed method.

The estimation of uncertainty followed the guidelines outlined in the Eurachem Guide for Quantifying Analytical Measurement Uncertainty [38], utilizing validation data from two fortification levels. Key sources contributing significantly to the overall uncertainty—such as the standard uncertainty from control solutions, uncertainty from linearity (coefficient of variation), precision of repeatability of control solutions, bias-related uncertainty, and precision of sample reproducibility—were prioritized in calculating the combined uncertainty.

2.6. Tissue Preparation and Histopathological Examination

The tissue and organ samples from 10 broiler chickens that had recently died (including the kidney, liver, thymus, bursa of Fabricius, and spleen) were fixed in buffered 10% formaldehyde for 24 h, followed by dehydration in (1) 70% ethylic alcohol for 60 min, (2) 95% ethylic alcohol for 45 min, and (3) absolute ethylic alcohol for 2 h. The clearing phase of the samples was made by repeated xylene immersions, followed by paraffin wax infiltrations [39]. The samples were automatically processed with the tissue processor Gemini AS Epredia (Runcorn, UK) and paraffin embedding was carried out with modular tissue-embedding center Microm EC 350-1 (Thermo Fischer Scientific, Waltham, MA, USA). The resulting blocks were cut at 3 μm using the RM 125RTS (Leica, Wetzlar, Hesse, Germany) microtome. The sections were stained with hematoxylin and eosin (HE) by using the Gemini AS Epredia Slide Stainer (Runcorn, UK). The examination was performed with an Olympus BX43 microscope coupled to an Olympus DP73 video camera and using the Olympus Cell^B analysis system version 3.4 (Shinjuku, Japan) for the microscopic evaluation of the samples.

3. Results

3.1. OTA Method Validation Parameters and Sample Contamination

The validation results of the OTA analytical method demonstrate that the developed approach effectively separates OTA from matrix interferences, achieving high selectivity primarily due to immunoaffinity columns that significantly reduce matrix effects. The UPLC-FLD chromatograms for the OTA reference standard solutions (3 ng/mL) demonstrate excellent resolution and similar clarity, as observed in the chromatograms for quality control (QC) maize materials and samples (Figure 1).

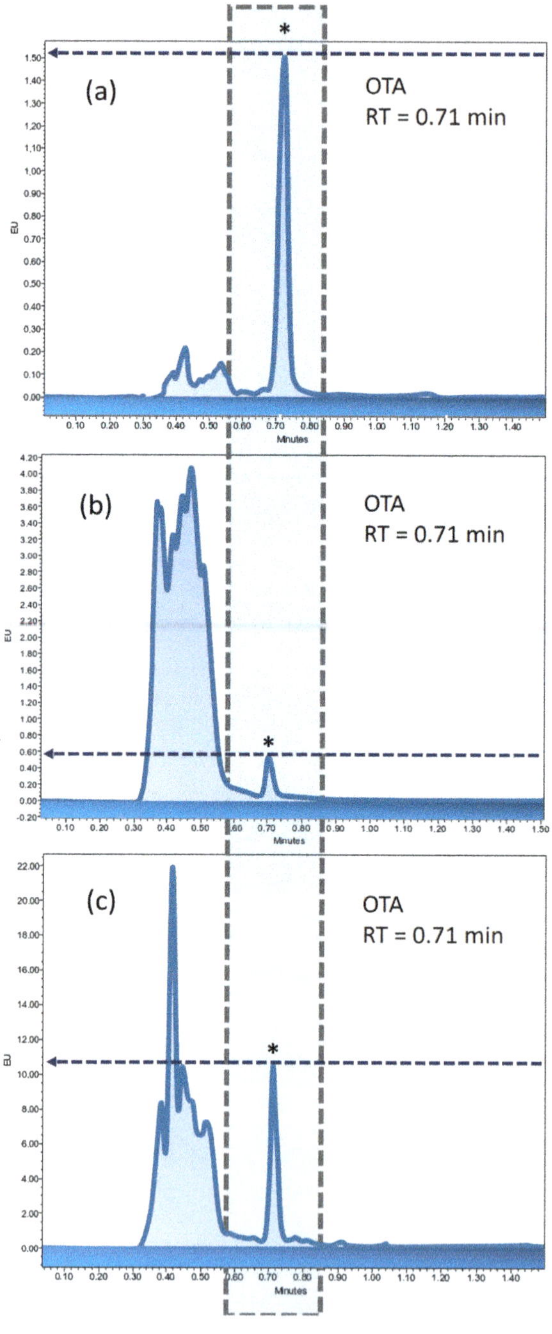

Figure 1. Typical UPLC-FLD chromatograms of OTA: (**a**) reference standard solution (3 ng/mL), (**b**) QC (maize), and (**c**) maize sample (M2). * represents the OTA peak.

Under optimal conditions, the calibration curve was established using five concentration levels of the OTA reference material solution, with two replicates at each level. The linearity was confirmed within the range of 0.5–10 ng/mL, with a regression equation of y = 7436.2x + 152.69 and a correlation coefficient (R^2) of 0.999, as per ISO 8466-1:2021 standards [40]. ANOVA analysis determined a standard error of 440.40 and a method standard deviation of 0.06, resulting in a variation coefficient of 1.37%, well below the generally accepted threshold of 5%. To evaluate accuracy, a 3 ng/mL solution from a different batch of certified OTA reference material was prepared and injected over 10 consecutive days, with two repetitions per injection. The accuracy of the method was clearly demonstrated, with results ranging from 95.83% to 104.65%.

LOQ is a crucial parameter, especially when evaluating contaminants with defined maximum levels. In this study, the LOQ was experimentally determined using trueness and precision criteria, based on maize QC material. To assess repeatability (RSD_r), six QC samples were analyzed in a single day, resulting in an average concentration of 0.95 µg/kg and an average recovery rate of 116.78%. The RSD_r was calculated at 7.95%, indicating excellent repeatability at low OTA concentrations. The within-laboratory reproducibility (RSD_{wR}) for the LOQ was 12.72%, which meets the requirements set by Commission Regulation (EU) No. 2782/2023, amending Regulation (EC) No. 401/2006 [36]. This falls well within the accepted RSD_r threshold of $\leq$20%, as well as the criteria specified in the EURL-MP guidelines [35], which recommend an LOQ value lower than 50 µg/kg.

From the nine maize samples tested, three samples were below the limit of quantification (LOQ)—two from Lunca, Buzău County, and one from Fierbinți. Additionally, three samples had OTA concentrations lower than 5 µg/kg: one from Moara Domnească at 4.88 µg/kg, and two from Nuci, Ilfov County, with levels of 1.47 µg/kg and 1.77 µg/kg, respectively. The sample collected from Coșereni contained 64.60 µg/kg of OTA, while another sample from Fierbinți showed a concentration of 14.68 µg/kg.

Notably, one maize sample (M3) from Nuci, Ilfov County, recorded a concentration of 228.75 µg/kg (Table 2), which is close to the maximum recommended OTA limit of 0.25 mg/kg established by Commission Recommendation No. 2016/1319 [41]. Consequently, poultry samples were collected from this farm to look into the possible impact of OTA-contaminated feed on kidney health. The OTA concentration in the feed mix was calculated using the fact that maize accounted for 60% of the diet specifically formulated for the chickens' growth phase. Among the nine samples tested, sample M3 exhibited a concerning concentration of 137.25 µg/kg, while sample M7 contained 38.76 µg/kg, warranting significant attention. Furthermore, sample M9 showed a moderate contamination level of 8.81 µg/kg.

Table 2. Ochratoxin A (OTA) contamination levels in maize samples from selected farms.

Sample Code	OTA in Maize * (µg/kg)	OTA in Feed Mix ** (µg/kg)
M1	4.88	2.93
M2	1.77	1.06
M3	228.75	137.25
M4	1.47	0.88
M5	<LOQ	-
M6	<LOQ	-
M7	64.60	38.76
M8	<LOQ	-
M9	14.68	8.81

Legend: <LOQ—below limit of quantification (0.810 µg/kg). * OTA in maize was analytically determined using UPLC-FLD. ** OTA in feed mix was calculated based on percentage of maize in maize-based diet formulated specifically for chickens' growth phase.

3.2. Gross and Histological Sample Result

During the necropsy of chickens from the farm located in Nuci, Ilfov County, where elevated levels of OTA (228.75 μg/kg) were detected in the contaminated maize, several pathological changes were observed. The kidneys of all individuals showed lesions marked by enlargement, degeneration, pallor, and increased firmness, with chalky deposits accumulating in the ureters (Figure 2). The serous membranes exhibited chalky white, fine, dry deposits. The liver showed hepatomegaly with a yellow-green coloration, characteristic of hepatic steatosis. Numerous foci of necrosis were seen on the surface of the liver, and on section, the liver tissue was soft and friable, suggesting massive degeneration of hepatocytes. Lymphoid organs such as the spleen, bursa of Fabricius, and thymus were reduced in size and pale.

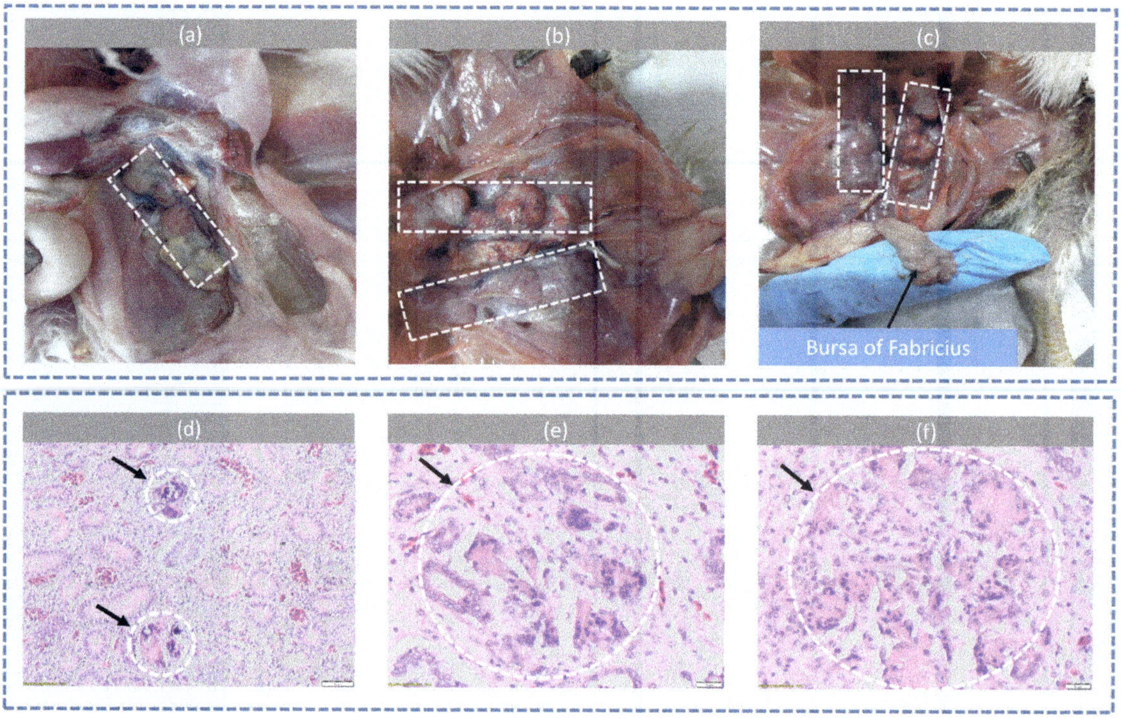

Figure 2. Gross and histopathological lesions from Ross 308 gout-affected commercial broiler chickens after exposure to feed with mycotoxins. (**a**) Swollen kidneys full of chalky deposit and the dilatation of ureters; the kidneys are marked by dashed white-line box. (**b**) The kidneys are enlarged, degenerate, and filled with urates, with bilateral multifocal petechiae; the kidneys are marked by the dashed white-line box. (**c**) The gross examination of the bursa of Fabricius reveals significant atrophy, with a notable reduction in size and weight, and pale discoloration; the kidneys are marked by the dashed white-line box. (**d**) Visceral gout, observed in the transverse cross-section of the chicken kidney. Bluish-purple urate deposits in the lumen of renal tubules forming urate cylinders (arrows), HE stain. (**e**) Gout tophi (indicated by a white dashed circle and arrow), with cellular spicules surrounded by macrophages and multinucleated giant cells, moderate interstitial lymphocytic and histiocytic nephritis, tubular degeneration, and necrosis in the chicken kidney, HE stain; (**f**) Gout tophi (indicated by a white dashed circle and arrow), with interstitial fibrosis and urate deposits in the chicken kidney, HE stain.

Microscopic examination revealed multifocal and random deposits in the kidneys, including colorless to intensely basophilic, spheroid, or sharp acicular crystalline deposits (urate tophi). Surrounding these deposits there were a moderate number of inflammatory cells, predominantly mononuclear lymphocytes, active macrophages, multinucleated giant cells, and heterophils mixed with eosinophilic and karyolytic cellular debris. Extravasated red blood cells extended into the adjacent interstitium. Renal tubules exhibited mild to moderate ectasia with a degenerate epithelium, enlarged epithelial cells with indistinct borders and microvacuolated cytoplasm, hypereosinophilic epithelial cytoplasm, and pyknotic nuclei. The interstitium contained infiltrates of lymphocytes and macrophages, along with signs of tissue fibrosis.

In the liver, hepatocytes displayed moderately increased cell volume with finely granular cytoplasm, indicative of extralobular hepatic glycogenosis. Some hepatocytes contained microvacuoles or macrovacuoles (steatosis), accompanied by perivascular, portobiliary, and sinusoidal inflammatory infiltrates. Multifocal hepatocytic necrosis was also noted. The intestinal mucosa was swollen and exhibited multiple hemorrhagic spots. In certain segments, the mucosa was necrotic and covered by a fibrinous exudate. Occasionally, the glandular body (proventriculus) showed erosions and ulcerations. A reduction in lymphocyte populations was observed in the lymphoid organs. The bursa of Fabricius frequently showed hypocellular lymphoid follicles with moderate to severe lymphocyte depletion and signs of cellular apoptosis.

4. Discussion

The UPLC-FLD technique serves as an affordable and user-friendly alternative to mass spectrometry (MS) or tandem mass spectrometry (MS/MS), while still delivering high specificity and sensitivity. This makes it particularly effective for detecting mycotoxins in various matrices, including maize, and suitable for multi-mycotoxin analyses. Recently, multiresidue immunoaffinity columns (IACs) have been developed to streamline the simultaneous purification of multiple mycotoxins, including aflatoxins B1, B2, G1, G2, OTA, and zearalenone [42]. Although these columns improve workflow efficiency, their recovery rates tend to be lower compared to single-target columns. OTA-specific immunoaffinity columns, however, remain highly effective for analyzing complex feed matrices, offering superior selectivity and enhanced analytical reliability. In addition, alternative methods such as QuEChERS combined with solid-phase extraction (SPE) have been explored as substitutes for IAC; however, they do not match the overall effectiveness. Ultimately, despite the higher cost associated with IAC, it continues to be regarded as the most effective and environmentally friendly option for mycotoxin analysis [43].

In our study, the UPLC-FLD method was validated for precision and accuracy, effectively separating OTA from matrix interferences while demonstrating high selectivity through the use of immunoaffinity columns. The method also enabled rapid analysis, with OTA showing a retention time of just 0.71 min and a total instrumental analysis time of 3 min. The UPLC chromatograms exhibited excellent resolution, high linearity ($R^2 = 0.999$), and accuracy ranging from 95.83% to 104.65%.

Leite et al. (2023) [44] developed and optimized a sensitive, precise, and robust multi-mycotoxin analytical method using ultra-high-performance liquid chromatography coupled with tandem mass spectrometry (UHPLC-MS/MS) for the detection of 23 regulated, non-regulated, and emerging mycotoxins in maize grain, including ochratoxin A (OTA). This approach used QuEChERS extraction with a C18 sorbent, followed by UHPLC-MS/MS analysis to accurately quantify these mycotoxins in maize samples. By analyzing blank samples and using a signal-to-noise ratio of 10:1, Leite et al. (2023) [44] achieved a limit of quantification (LOQ) of 3.7 μg/kg, with recovery rates between 75.8% and 98.1%. In comparison, our study achieved a significantly lower LOQ of 0.81 μg/kg, along with a recovery rate of 116.78%.

Another key validation parameter, RSD_{wR}, which influences the standard uncertainty of the method, was assessed in this study. In our research, we observed an RSD_{wR} of 12.72% at 0.81 µg/kg, while Silva et al. (2019) [45] reported a slightly higher RSD_{wR} of 14.5% at a concentration of 1.5 µg/kg. Leite et al. (2023) [44] recorded an RSD_{wR} of 13.6% in a poultry feed analysis spiked at 2 µg/kg, further underscoring the influence of feed matrices on analyte recovery rates.

Romania's warm and humid climate, especially in regions such as the Southern Plain and Dobrogea, creates optimal conditions for *Aspergillus* species, key OTA producers. These conditions increase the risk of OTA contamination, particularly during maize storage in silos, posing potential health risks if not effectively managed. Between January 2012 and July 2015, post-harvest OTA contamination in Romanian maize was reported at an incidence rate of 6.8%, with 4 out of 59 samples testing positive (Gagiu et al., 2018) [46]. OTA levels in contaminated samples ranged from below 2.50 µg/kg to a maximum of 6.72 µg/kg, with a mean of 2.70 ± 0.43 µg/kg and a median below 2.50 µg/kg. Of the samples, 6.8% exceeded the maximum allowable OTA level, raising food safety concerns in affected areas. Processed maize products showed the highest OTA levels, with maize grains stored in silos reaching up to 3.39 µg/kg in 2012 and corn germ samples reaching 5.64–6.72 µg/kg in 2014.

Maize, however, is not the only feed susceptible to OTA contamination. Alexa et al. (2013) [47] found a relatively low OTA incidence in 2010 wheat samples in western Romania, with a maximum contamination level of 11.3 µg/kg. However, OTA incidence increased significantly in 2011, with 92.31% of wheat samples testing positive and contamination levels reaching up to 25.70 µg/kg, demonstrating annual fluctuations in contamination across feed types. OTA contamination has also been reported in maize from other regions, indicating similar risks in various climates. For instance, Kortei et al. (2021) [48] assessed OTA levels in Ghanaian maize, collecting samples from village markets from July to December 2020. Of the 180 samples analyzed, 103 contained OTA, with contamination levels ranging from 0 to 97.51 µg/kg. Of these, 52.22% exceeded the EFSA tolerable limit, while 49.44% exceeded Ghana's safety standard.

In Doha, Qatar, Alkuwari et al. (2021) [49] reported that all maize samples (n = 13) collected from local markets tested positive for OTA, with levels between 1.55 and 2.58 µg/kg. Similarly, Beg et al. (2006) [50] investigated OTA contamination in yellow maize and other feed ingredients in a Kuwaiti poultry feed production facility. Among the 14 broiler starter feed samples, 93% contained less than 5 µg/kg OTA, and 32 maize samples analyzed showed that half had OTA concentrations below this threshold.

The nephrotoxic effects of ochratoxins have been investigated for a long time. Grossly, these lesions include enlarged kidneys with urate retention. Microscopically, renal tubular degeneration and necrosis, along with cast formation and urate tophi, are typically observed. Lymphoid organ suppression (lymphocyte depletion) and hepatic necrosis may also occur. Similar results have been reported by other studies [51]. Santin et al. (2002) [52] conducted a necropsy analysis of chickens fed a diet containing OTA at 2 mg/kg for three to four weeks, observing alterations in internal organs such as pale, swollen kidneys and enlarged, yellowish, friable livers. Histopathological examinations revealed the vacuolation and megalocytosis of hepatocytes, hyperplasia of the biliary epithelium, and hypertrophy of renal proximal tubular epithelial cells. In another study, broiler chickens fed a diet with 2 mg/kg of OTA showed no changes within the first three weeks of treatment [53]. However, significant changes were noted after five weeks, including kidney swelling, slight liver enlargement, and a reduction in the size of the bursa of Fabricius. Microscopic analyses indicated marked changes in lesion scores, with the highest scores observed in the kidneys, followed by the liver, bursa, spleen, and thymus. Kumar et al. (2004) [54] confirmed that OTA is more nephrotoxic than hepatotoxic for broilers, with the most significant effects seen in animals fed a diet contaminated with OTA at 2 mg/kg. These effects included the atrophy of the bursa, thymus, and spleen, along with lymphocyte depletion. In a recent study on two commercial poultry farms, Bozzo et al. (2008) [55] detected OTA in all feed samples, with concentrations ranging from 0.160 to 0.332 mg/kg. The OTA-contaminated feed was administered to the animals for at least two months. Postmortem inspections, along with cytological and histological examinations of layer hens, evidenced gross and microscopic lesions in the kidneys and liver. Sawale et al. (2009) [56] reported negative effects of OTA on performance, hematobiochemical disturbances, and severe immunosuppression in laying hens fed a diet contaminated with OTA at 1 mg/kg for 60 days. Our study provides additional evidence in this field.

5. Conclusions

This study presents a developed and validated UPLC-FLD method that exhibits robust precision and accuracy in the detection of OTA in maize. The method effectively separates OTA from matrix interferences, thereby ensuring reliable detection at low concentrations. The maize samples tested from various counties in Romania reveal significant variability in OTA concentrations. Notably, maize sample M3 exhibited a concentration of 228.75 μg/kg, which, while below the maximum recommended limit of 250 μg/kg, underscores the potential risk associated with OTA contamination in maize. Histopathological assessments of the collected poultry samples indicated nephrological damage that may be associated with OTA presence, including kidney enlargement, urate deposits, and tubular degeneration, along with lymphoid organ suppression and hepatic necrosis. These findings corroborate existing research on OTA's severe impact on poultry health. This study underscores the necessity for stringent feed testing and management practices to address OTA contamination and mitigate its adverse effects. Furthermore, the analysis highlights the effectiveness of the UPLC-FLD method in providing the accurate quantification of mycotoxins, with high correlation coefficients and recovery rates affirming its reliability. This research emphasizes the critical importance of continuous mycotoxin surveillance to ensure feed safety and protect both animal and human health, offering valuable insights for the agricultural and public health sectors.

Author Contributions: Conceptualization, S.-I.B., V.A.I., E.G. and L.B.; formal analysis, V.A.I.; investigation, S.-I.B., V.A.I., E.G., O.-C.B., E.Ş.I., A.B., E.P. and V.E.B.; methodology, S.-I.B., V.A.I., E.G., O.-C.B., E.Ş.I., A.B., E.P., V.E.B. and L.B.; supervision, S.-I.B. and V.A.I.; validation, V.A.I. and E.G.; writing—original draft, V.A.I., E.G., O.-C.B., E.Ş.I., A.B. and L.B.; writing—review and editing, S.-I.B., V.A.I., E.G., O.-C.B., E.Ş.I., A.B., E.P., V.E.B. and L.B. All authors have read and agreed to the published version of the manuscript.

Funding: This work was supported by a grant from the University of Agronomic Sciences and Veterinary Medicine of Bucharest, project number 2022-0017/16.05.2022, acronym ToxiSafe, within IPC 2021.

Institutional Review Board Statement: Not applicable.

Informed Consent Statement: Not applicable.

Data Availability Statement: Data are available on request from the authors.

Conflicts of Interest: The authors declare no conflicts of interest.

References

1. Miljkovic, A.; Mantle, P. Chromatographic Fractionation of *Penicillium polonicum* Fermentation Metabolites in Search of the Nephrotoxin(s) for Rats. *Life* **2022**, *12*, 747. [CrossRef] [PubMed]
2. Rahim, H.M.; Othman, D.H.; Majeed, R.K.; Saeed, N.M.; Ismaeel, D.O. Mycotoxins in poultry feed and feed ingredients in Sulaymaniyah, kurdistan region of Iraq. *Bas. J. Vet. Res.* **2020**, *19*, 210–222. [CrossRef]
3. Ismaiel, A.; Papenbrock, J. Mycotoxins: Producing fungi and mechanisms of phytotoxicity. *Agriculture* **2015**, *5*, 492–537. [CrossRef]
4. Alshannaq, A.; Yu, J.-H. Occurrence, toxicity, and analysis of major mycotoxins in food. *Int. J. Environ. Res. Public Health* **2017**, *14*, 632. [CrossRef] [PubMed]
5. Zhai, S.; Zhu, Y.; Feng, P.; Li, M.; Wang, W.; Yang, L.; Yang, Y. Ochratoxin A: Its Impact on Poultry Gut Health and Microbiota, an Overview. *Poult. Sci.* **2021**, *100*, 101037. [CrossRef]

6. Mohamed, R.S.; Attia, A.I.; El-Mekkawy, M.M.; Ismail, F.S.; Salah, A.S.; Nicotra, M.; Di Cerbo, A.; Azzam, M.M.; Alagawany, M. Leverage of *Matricaria chamomilla* L. Oil Supplementation over Ochratoxin A in Growing Quails. *J. Food Qual.* **2024**, *2024*, 7254066. [CrossRef]
7. Corbu, V.M.; Georgescu, A.M.; Gheorghe-Barbu, I.; Csutak, O. Antifungal activity of microbial strains isolated from Romanian spontaneously fermented dairy products. *Agrolife Sci. J.* **2024**, *13*, 54–64.
8. Bilal, R.M.; Tahir, M.A.; Shahzad, A.; Farag, M.R.; Siddiq, A.; El-Shall, N.A.; Dhama, K.; Elnesr, S.S.; Alagawany, M. Yeast and Derived Products: Their Uses in Preventing Mycotoxins in Poultry Feeds. *World's Poult. Sci. J.* **2023**, *79*, 351–378. [CrossRef]
9. Beia, I.S.; Ciceoi, R.; Beia, V.E.; Bădulescu, L.; Smedescu, D.I.; Fîntîneru, A. Mycotoxins—Incidence, impact on feed, food safety, food chain and economic losses. *Sci. Pap. Ser. Manag. Econ. Eng. Agric. Rural Dev.* **2023**, *23*, 85–94.
10. Eskola, M.; Kos, G.; Elliott, C.T.; Hajšlová, J.; Mayar, S.; Krska, R. Worldwide Contamination of Food-Crops with Mycotoxins: Validity of the Widely Cited 'FAO Estimate' of 25%. *Crit. Rev. Food Sci. Nutr.* **2020**, *60*, 2773–2789. [CrossRef]
11. Denli, M. Implications of Mycotoxins in Livestock Feeds. *Agrolife Sci. J.* **2015**, *4*, 52–55.
12. Sîrbu, V.I.; Popa (Burlacu), A.; Israel-Roming, F. Mycotoxins in feed: An overview on biological effects and decontamination methods. *Agrolife Sci. J.* **2020**, *9*, 285–296.
13. Lee, H.J.; Kim, H.D.; Ryu, D. Practical Strategies to Reduce Ochratoxin A in Foods. *Toxins* **2024**, *16*, 58. [CrossRef] [PubMed]
14. Beia, I.S.; Ciceoi, R.; Ion, V.A.; Tănăsuică, R.; Bădulescu, L. Management of mycotoxins contamination of feed inputs on the agri-food chain. *Sci. Pap. Ser. Manag. Econ. Eng. Agric. Rural Dev.* **2022**, *22*, 109–116.
15. Kőszegi, T.; Poór, M. Ochratoxin a: Molecular Interactions, Mechanisms of Toxicity and Prevention at the Molecular Level. *Toxins* **2016**, *8*, 111. [CrossRef] [PubMed]
16. Colombo, R.; Papetti, A. Pre-Concentration and Analysis of Mycotoxins in Food Samples by Capillary Electrophoresis. *Molecules* **2020**, *25*, 3441. [CrossRef]
17. Kirinčič, S.; Škrjanc, B.; Kos, N.; Kozolc, B.; Pirnat, N.; Tavčar-Kalcher, G. Mycotoxins in Cereals and Cereal Products in Slovenia—Official Control of Foods in the Years 2008–2012. *Food Control.* **2015**, *50*, 157–165. [CrossRef]
18. Ahuja, V.; Singh, A.; Paul, D.; Dasgupta, D.; Urajová, P.; Ghosh, S.; Singh, R.; Sahoo, G.; Ewe, D.; Saurav, K. Recent Advances in the Detection of Food Toxins Using Mass Spectrometry. *Chem. Res. Toxicol.* **2023**, *36*, 1834–1863. [CrossRef]
19. Widiyanti, P.M.; Maryam, R. Ochratoxin A (OTA) Contamination Detection by Enzyme-Linked Immunosorbent Assay (ELISA). *IOP Conf. Ser. Earth Environ. Sci.* **2023**, *1253*, 012100. [CrossRef]
20. Kuruc, J.; Hegstad, J.; Jung Lee, H.; Simons, K.; Ryu, D.; Wolf-Hall, C. Infestation and Quantification of Ochratoxigenic Fungi in Barley and Wheat Naturally Contaminated with Ochratoxin a. *J. Food Prot.* **2015**, *78*, 1350–1356. [CrossRef]
21. Zhao, Y.; Chen, D.; Duan, H.; Li, P.; Wu, W.; Wang, X.; Poapolathep, A.; Poapolathep, S.; Logrieco, A.F.; Pascale, M.; et al. Sample Preparation and Mass Spectrometry for Determining Mycotoxins, Hazardous Fungi, and Their Metabolites in the Environment, Food, and Healthcare. *Trends Anal. Chem.* **2023**, *160*, 116962. [CrossRef]
22. Arroyo-Manzanares, N.; Campillo, N.; López-García, I.; Hernández-Córdoba, M.; Viñas, P. High-Resolution Mass Spectrometry for the Determination of Mycotoxins in Biological Samples. A Review. *Microchem. J.* **2021**, *166*, 106197. [CrossRef]
23. García-Nicolás, M.; Arroyo-Manzanares, N.; Campillo, N.; Viñas, P. Cellulose-Ferrite Nanocomposite for Monitoring Enniatins and Beauvericins in Paprika by Liquid Chromatography and High-Resolution Mass Spectrometry. *Talanta* **2021**, *226*, 122144. [CrossRef] [PubMed]
24. Alvito, P.; Assunco, R.M.; Bajard, L.; Martins, C.; Mengelers, M.J.B.; Mol, H.; Namorado, S.; van den Brand, A.D.; Vasco, E.; Viegas, S.; et al. Current Advances, Research Needs and Gaps in Mycotoxins Biomonitoring under the HBM4EU—Lessons Learned and Future Trends. *Toxins* **2022**, *14*, 826. [CrossRef] [PubMed]
25. Viegas, S.; Veiga, L.; Malta-Vacas, J.; Sabino, R.; Figueredo, P.; Almeida, A.; Viegas, C.; Carolino, E. Occupational Exposure to Aflatoxin (AFB 1) in Poultry Production. *J. Toxicol. Environ. Health A* **2012**, *75*, 1330–1340. [CrossRef]
26. Viegas, S.; Assunção, R.; Martins, C.; Nunes, C.; Osteresch, B.; Twarużek, M.; Kosicki, R.; Grajewski, J.; Ribeiro, E.; Viegas, C. Occupational Exposure to Mycotoxins in Swine Production: Environmental and Biological Monitoring Approaches. *Toxins* **2019**, *11*, 78. [CrossRef]
27. Muñoz-Solano, B.; Lizarraga Pérez, E.; González-Peñas, E. Monitoring Mycotoxin Exposure in Food-Producing Animals (Cattle, Pig, Poultry, and Sheep). *Toxins* **2024**, *16*, 218. [CrossRef]
28. Tsiouris, V.; Tassis, P.; Raj, J.; Mantzios, T.; Kiskinis, K.; Vasiljević, M.; Delić, N.; Petridou, E.; Brellou, G.D.; Polizopoulou, Z.; et al. Investigation of a Novel Multicomponent Mycotoxin Detoxifying Agent in Amelioration of Mycotoxicosis Induced by Aflatoxin-B1 and Ochratoxin a in Broiler Chicks. *Toxins* **2021**, *13*, 367. [CrossRef]
29. Awais, M.M.; Mehtab, U.; Anwar, M.I.; Hameed, M.R.; Akhtar, M.; Raza, A.; Aisha, R.; Muhammad, F.; Saleemi, M.K.; Fayyaz, A. Mitigation potential of individual and combined dietary supplementation of local Bentonite Clay and Distillery Sludge against Ochratoxin-A induced toxicity in broilers. *BMC Vet. Res.* **2022**, *18*, 375. [CrossRef]
30. Solcan, C.; Pavel, G.; Floristean, V.; Chiriac, I.; Lencu, B.; Solcan, G. Effect of ochratoxin A on the intestinal mucosa and mucosa-associated lymphoid tissues in broiler chickens. *Acta Vet. Hung.* **2015**, *63*, 30–48. [CrossRef]
31. Tahir, M.A.; Abbas, A.; Muneeb, M.; Bilal, R.M.; Hussain, K.; Abdel-Moneim, A.-M.E.; Farag, M.R.; Dhama, K.; Elnesr, S.S.; Alagawany, M. Ochratoxicosis in Poultry: Occurrence, Environmental Factors, Pathological Alterations and Amelioration Strategies. *World's Poult. Sci. J.* **2022**, *78*, 727–749. [CrossRef]

32. Khatoon, A.; Khan, M.Z.; Khan, A.; Javed, I. Toxicopathological and Serum Biochemical Alterations Induced by Ochratoxin A in Broiler Chicks and Their Amelioration by Locally Available Bentonite Clay. *Pak. J. Agric. Sci.* **2016**, *53*, 977–984. [CrossRef]
33. Schrenk, D.; Bignami, M.; Bodin, L.; Chipman, J.K.; del Mazo, J.; Grasl-Kraupp, B.; Hogstrand, C.; Hoogenboom, L.R.; Leblanc, J.-C.; Nebbia, C.S.; et al. Risks for animal health related to the presence of ochratoxin A (OTA) in feed. *EFSA J.* **2023**, *21*, 8375. [CrossRef]
34. Liu, M.; Zhao, L.; Gong, G.; Zhang, L.; Shi, L.; Dai, J.; Han, Y.; Wu, Y.; Khalil, M.M.; Sun, L. Invited review: Remediation strategies for mycotoxin control in feed. *J. Anim. Sci. Biotechnol.* **2022**, *13*, 19. [CrossRef]
35. EURL-MP-Background doc_003 Guidance Document Performance Criteria v1.4 Draft 17.05.2024-WUR. Available online: https://www.wur.nl/en/show/EURL-MP-background-doc_003-Guidance-document-performance-criteria-v1.1-draft-17.09.2021.htm (accessed on 9 September 2024).
36. Commission Implementing Regulation (EU) 2023/2782, of 14 December 2023. Laying down the Methods of Sampling and Analysis for the Control of the Levels of Mycotoxins in Food and Repealing Regulation (EC) No 401/2006. Available online: https://eur-lex.europa.eu/legal-content/EN/TXT/?uri=CELEX:02023R2782-20240324 (accessed on 18 August 2023).
37. OtaCLEAN™ for Ochratoxin A. Available online: https://www.lctech.de/en/products/mycotoxin-products/otaclean (accessed on 18 August 2023).
38. Ellison, S.L.R.; Williams, A. (Eds.) *Eurachem/CITAC Guide: Quantifying Uncertainty in Analytical Measurement*, 3rd ed.; EURACHEM: Teddington, UK, 2012; ISBN 978-0-948926-30-3. Available online: https://www.eurachem.org (accessed on 18 August 2023).
39. Dinescu, G.; Gagniuc, E.; Tomescu, G. Complex investigations in diagnosis of marek's disease in poultry. *Sci. Works. Ser. C. Vet. Med.* **2023**, *2*, 55–61.
40. *ISO 8466-1:2021*; Water Quality—Calibration and Evaluation of Analytical Methods, Part 1: Linear Calibration Function Published (Edition 2, 2021). International Organization for Standardization: Geneva, Switzerland, 2021.
41. Commission Recommendation (EU) 2016/1319 of 29 July 2016 Amending Recommendation 2016/576/EC as Regards Deoxynivalenol, Zearalenone, Ochratoxin A in Pet Food. Available online: https://eur-lex.europa.eu/legal-content/EN/TXT/?uri=CELEX:32016H1319 (accessed on 18 August 2023).
42. Muñoz-Solano, B.; González-Peñas, E. Mycotoxin Determination in Animal Feed: An LC-FLD Method for Simultaneous Quantification of Aflatoxins, Ochratoxins and Zearelanone in This Matrix. *Toxins* **2020**, *12*, 374. [CrossRef]
43. Mackay, N.; Marley, E.; Leeman, D.; Poplawski, C.; Donnelly, C. Analysis of Aflatoxins, Fumonisins, Deoxynivalenol, Ochratoxin A, Zearalenone, HT-2, and T-2 Toxins in Animal Feed by LC–MS/MS Using Cleanup with a Multi-Antibody Immunoaffinity Column. *J. Aoac Int.* **2022**, *105*, 1330–1340. [CrossRef]
44. Leite, M.; Freitas, A.; Barbosa, J.; Ramos, F. Comprehensive assessment of different extraction methodologies for optimization and validation of an analytical multi-method for determination of emerging and regulated mycotoxins in maize by UHPLC-MS/MS. *Food Chem. Adv.* **2023**, *2*, 100145. [CrossRef]
45. Silva, A.S.; Brites, C.; Pouca, A.V.; Barbosa, J.; Freitas, A. UHPLC-ToF-MS method for determination of multi-mycotoxins in maize: Development and validation. *Curr. Res. Food Sci.* **2019**, *1*, 1–7. [CrossRef]
46. Gagiu, V.; Mateescu, E.; Armeanu, I.; Dobre, A.A.; Smeu, I.; Cucu, M.E.; Oprea, O.A.; Iorga, E.; Belc, N. Post-Harvest Contamination with Mycotoxins in the Context of the Geographic and Agroclimatic Conditions in Romania. *Toxins* **2018**, *10*, 533. [CrossRef]
47. Alexa, E.; Dehelean, C.A.; Poiana, M.A.; Radulov, I.; Cimpean, A.M.; Bordean, D.M.; Tulcan, C.; Pop, G. The occurrence of mycotoxins in wheat from western Romania and histopathological impact as effect of feed intake. *Chem. Cent. J.* **2013**, *7*, 99. [CrossRef] [PubMed]
48. Kortei, N.K.; Annan, T.; Kyei-Baffour, V.; Essuman, E.K.; Okyere, H.; Tettey, C.O. Exposure and risk characterizations of ochratoxins A and aflatoxins through maize (*Zea mays*) consumed in different agro-ecological zones of Ghana. *Sci. Rep.* **2021**, *11*, 2339. [CrossRef] [PubMed]
49. Alkuwari, A.; Hassan, Z.U.; Zeidan, R.; Al-Thani, R.; Jaoua, S. Occurrence of Mycotoxins and Toxigenic Fungi in Cereals and Application of Yeast Volatiles for Their Biological Control. *Toxins* **2022**, *14*, 404. [CrossRef] [PubMed]
50. Beg, M.U.; Al-Mutairi, M.; Beg, K.R.; Al-Mazeedi, H.M.; Ali, L.N.; Saeed, T. Mycotoxins in Poultry Feed in Kuwait Arch. *Environ. Contam. Toxicol.* **2006**, *50*, 594–602. [CrossRef] [PubMed]
51. Mehtab, U.; Tahir, M.A.; Abbas, R.Z.; Abbas, A.; Hussain, K.; Siddiqu, F.; Mohsin, M.; Rani, Z.; Rehman, A.; Yasin, R. Ochratoxin A occurrence, its pathological effects on poultry health and decontamination approaches. *J. Hell. Vet. Medical. Soc.* **2022**, *72*, 3257–3262. [CrossRef]
52. Santin, E.; Paulillo, A.C.; Maiorka, P.C.; Alessi, A.C.; Krabbe, E.L.; Maiorka, A. The effects of ochratoxin/aluminosilicate interaction on the tissues and humoral immune response of broilers. *Avian Path.* **2002**, *31*, 73–79. [CrossRef] [PubMed]
53. Gupta, S.; Jindal, N.; Khokhar, R.S.; Asrani, R.K.; Ledoux, D.R.; Rottinghaus, G.E. Individual and combined effects of ochratoxin A and Salmonella enterica serovar Gallinarum infection on pathological changes in broiler chickens. *Avian Path.* **2008**, *37*, 265–272. [CrossRef] [PubMed]
54. Kumar, A.; Jindal, N.; Shukla, C.L.; Asrani, R.K.; Ledoux, D.R.; Rottinghaus, G.E. Pathological changes in broiler chickens fed ochratoxin A and inoculated with Escherichia coli. *Avian Path.* **2004**, *33*, 413–417. [CrossRef] [PubMed]

55. Bozzo, G.; Ceci, E.; Bonerba, E.; Desantis, S.; Tantillo, G. Ochratoxin A in laying hens: High-performance liquid chromatography detection and cytological and histological analysis of target tissues. *J. Appl. Poult. Res.* **2008**, *17*, 151–156. [CrossRef]
56. Sawale, G.K.; Gosh, R.C.; Ravikanth, K.; Maini, S.; Rekhe, D.S. Experimental Mycotoxicosis in Layer Induced by Ochratoxin A and its Amelioration with Herbomineral Toxin Binder 'Toxiroak'. *Int. J. Poult. Sci.* **2009**, *8*, 798–803. [CrossRef]

life

Case Report

Chronic Copper Bilysinate Poisoning in Five Texel Sheep: A Case Report

Dalma Pivariu [1], Adrian Nechita Oros [1], Alexandru Tabaran [2,*], Francesca Caloni [3], Pompei Bolfa [4] and Andras-Laszlo Nagy [4]

1 Department of Toxicology, Faculty of Veterinary Medicine, University of Agricultural Sciences and Veterinary Medicine Cluj-Napoca, Calea Manastur 3-5, 400372 Cluj-Napoca, Romania; dalma.pivariu@usamvcluj.ro (D.P.); orosnadrian@yahoo.com (A.N.O.)
2 Department of Pathology, Faculty of Veterinary Medicine, University of Agricultural Sciences and Veterinary Medicine Cluj-Napoca, Calea Manastur 3-5, 400372 Cluj-Napoca, Romania
3 Department of Environmental Science and Policy (ESP), Universita degli Studi di Milano, Via Celoria 10, 20133 Milan, Italy; francesca.caloni@unimi.it
4 Department of Biomedical Sciences, Ross University School of Veterinary Medicine, Basseterre P.O. Box 334, Saint Kitts and Nevis; pompeibolfa@gmail.com (P.B.); anagy@rossvet.edu.kn (A.-L.N.)
* Correspondence: alexandru.tabaran@usamvcluj.ro

Abstract: Copper is an essential trace element but becomes toxic in overexposed animals. Sheep are the domestic species most prone to chronic copper poisoning, as a slight increase in the dietary concentration can lead to liver accumulation and the development of clinical signs in this species. Common sources of copper in the diet are feed additives and mineral supplements, which are commonly used in pigs and poultry. Recently, new copper supplements were registered for animal nutrition, including copper bilysinate. This study describes an episode of presumed chronic copper poisoning in Five Texel sheep, which were exposed to a compound feed containing copper bilysinate. Four weeks after the introduction of the compound feed into the diet, the first animal started to show typical clinical signs of chronic copper poisoning and died, followed by another animal a week later. Despite removing the compound feed from the diet, a third sheep died 3 weeks later. Two animals survived and fully recovered. Necropsy and histology showed characteristic gross and microscopical lesions typical of copper poisoning. The case report highlights the potential toxic effect of copper bilysinate in sheep.

Keywords: hemolysis; trace minerals; potentially toxic element; toxicity

1. Introduction

Copper (Cu) is an essential trace element for both ruminants and monogastrics [1] and plays an important role in many physiological processes, as it is a component of proteins, enzymes, and nucleic acids [2–6]. As a component of essential enzymatic systems, copper is involved in energy production, immune response, growth, nervous system function, and connective tissue formation [1,5].

Copper toxicosis is more common in sheep than in cattle or monogastric animals [1], causing important economic losses [7]. Cattle and certain breeds of dogs, notably Bedlington Terriers, are also sensitive. In Bedlington Terriers, an inherited sensitivity to copper toxicosis similar to Wilson disease in humans has been identified. Chronic copper poisoning has been reported in other breeds of dogs, including Labrador Retrievers, West Highland White Terriers, Skye Terriers, Keeshonds, American Cocker Spaniels, and Doberman Pinschers [8,9]. Other species, such as pigs, poultry, and horses, may be affected but are more resistant [7].

The high susceptibility of sheep to copper toxicosis is due to their low dietary requirement and their inability to increase copper excretion in response to increased dietary intake [6]. Certain breeds of sheep are more sensitive than others, Texel sheep being considered one of the most sensitive breeds to copper toxicosis [6]. Milk sheep in intensive farming [7] and young animals are more prone to copper poisoning than adult ones [6]. This age-related susceptibility may be attributable to their higher copper absorption capacity than in adult sheep and their underdeveloped biliary excretion [10].

In sheep, when the diet contains elevated copper levels, or is deficient in molybdenum, copper accumulates in the liver [1]. The most important sources of copper in the diet are feed additives such as copper sulfate, copper chloride, and copper oxide [9]. The use of trace mineralized salt in sheep diets or forage harvested from fields fertilized with poultry or swine manure with increased copper are other causes of poisoning [7]. Misformulation of rations or errors in the mixing of feed can result in high concentrations of copper.

Chronic copper poisoning is a common global problem for the sheep industry, and outbreaks were reported and case reports were published in many sheep-rearing countries, such as Iran, Australia, New Zealand, South Africa, USA, Brazil, and Canada [7]. In Europe, recent reports from the United Kingdom described cases of copper poisoning in sheep, associated with organic farming and the consumption of red clover (*Trifolium pratense*) and white clover (*Trifolium repens*) [11]. It is also a common problem in the Greek sheep industry [11]. Cases were also reported in Spain, Turkey, and the Scandinavian countries [7]. In Belgium, commercially prepared milk replacers with high copper levels were involved in poisoning cases in calves [11].

In this case report, we describe an episode of accidental chronic copper poisoning in Texel sheep due to feeding the animals with a compound feed containing mineral supplements intended for pigs. The mineral supplement contained copper bilysinate, a new form of copper, which was authorized in 2014 in the European Union for animal nutrition for all animal species. Also in 2014, European Food Safety Authority (EFSA) showed an equivalent bioavailability between copper bilysinate and copper sulfate [12]. This feed additive can be a powder or a granulate with a content of copper $\geq$ 14.5% and lysine-HCl $\geq$ 84%; the active substance name is copper chelate of L-lysinate-HCl, and the chemical formula is $Cu(C_6H_{13}N_2O_2)_2$ [12]. In addition to the high copper content, according to the manufacturer, the commercial product involved in this poisoning case reduces copper excretion by 43.6% in pigs (the species for which the product is labeled) [13], which can potentially increase the risk for hepatic accumulation and chronic poisoning in sheep.

2. Case Description

The episode occurred in a small owner-operated unit in Cluj County, Romania, in January 2022. The farmer, who was new in sheep husbandry, incorrectly purchased feed with added copper bilysinate intended for pigs. Five animals were involved in this poisoning case: one male and four pregnant ewes, Texel breed of sheep. The animals were acquired one month before the onset of the clinical signs and were housed in a stable, where they had permanent free access to fresh water and were fed with good-quality hay and a compound feed with mineral supplement intended for pigs, containing 14.22 mg/kg copper bilysinate. There was no added molybdenum in this commercial product.

The animals started to show clinical signs 6 weeks after the introduction of the commercial compound feed into the diet. The first animal to show clinical signs was a three-year-old pregnant ewe. The animal showed anorexia, weakness, and apathy. Jaundice, characterized by the yellow discoloration of the conjunctiva and the skin, hemoglobinuria, abdominal pain, rumen stasis, and normal body temperature were observed. Although it was able to rise, the animal preferred sternal recumbency and died 2 days after the onset of the clinical signs. A week later, a second animal, a three-year-old male, started to show similar clinical signs and died. After this event, the owner suspected a feed-related problem and stopped feeding the surviving animals with the high copper level feed. The hay was examined, and no visible signs of impurities or toxic plants were identified. The animals had permanent free access to fresh water, the public water system being the source of water at this farm.

Although the compound feed was removed from the diet, after another 3 weeks (week 10 after the introduction of the compound feed), another pregnant ewe died, presenting similar clinical signs and lesions as the previous animals showed.

The two surviving pregnant ewes presented mild clinical signs of anorexia and apathy, fully recovered, and gave birth to healthy lambs.

Blood smear realized from peripheral blood showed signs of oxidative hemolytic anemia, with the presence of erythrocyte ghosts and Heinz bodies (Figure 1).

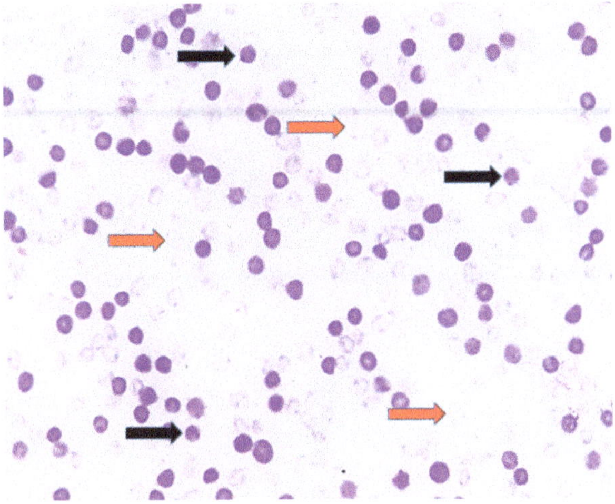

Figure 1. Blood smear, evidence of Heinz bodies (black arrows) and ghost erythrocytes (red arrows), indicating hemolytic anemia; Diff-Quick stain.

A complete necropsy was performed on the deceased animals, and the poisoning was confirmed through pathological findings. Grossly, the animals showed marked, diffuse icterus (Figure 2A,B), splenomegaly, pale, friable, and enlarged liver (Figure 3A), distended gallbladder, dark-brown discoloration of the renal cortex (Figure 4A), and hemoglobinuria (Figure 5).

The main histological findings were acute, massive, diffuse hepatic necrosis (Figure 3B,C) and diffuse, severe renal tubular necrosis associated with the presence of hemoglobin casts within renal tubules (Figure 4B,C).

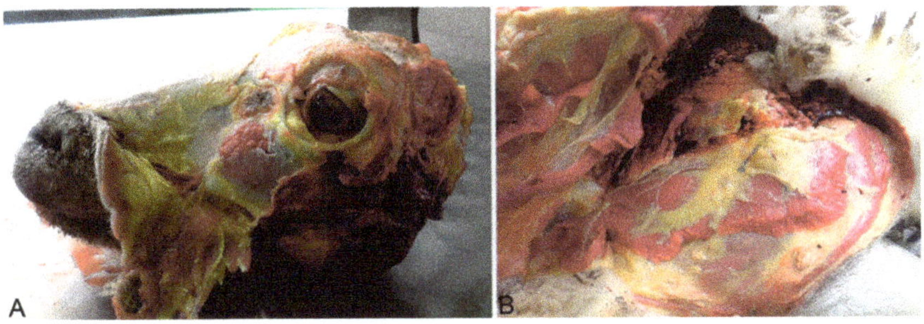

Figure 2. (**A**) (Head), (**B**) (Left hindlimb). Generalized icterus, yellowish discoloration of the skin and subcutaneous fat, caused by hyperbilirubinemia resulting from intravascular hemolysis.

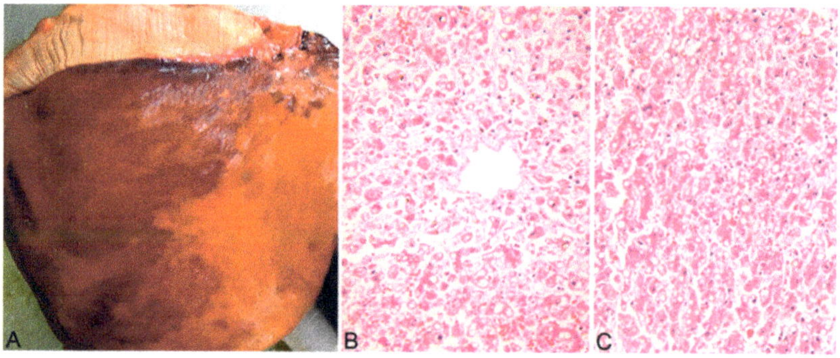

Figure 3. Pale, enlarged liver (**A**). Massive, diffuse hepatocellular necrosis, centrolobular area (**B**) and midzonal area (**C**). Hematoxylin-eosin stain (**B**,**C**).

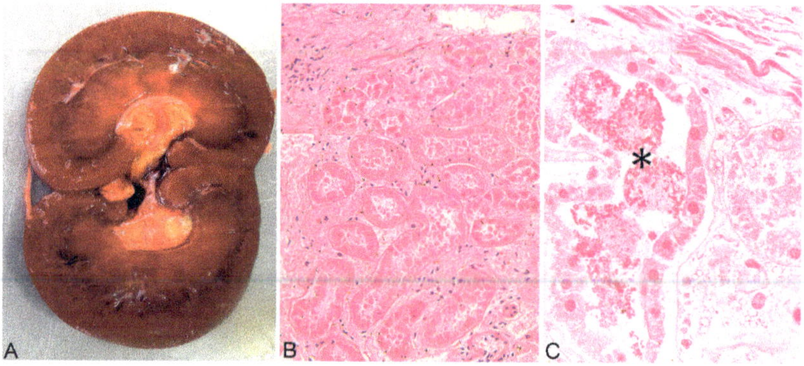

Figure 4. Kidney, diffuse dark-brown discoloration of the renal cortex (**A**), Acute, severe renal tubular necrosis—all tubular structures in the section are affected (**B**). Multifocal intraluminal hemoglobin casts—asterisk (**C**). Hematoxylin-eosin stain (**B**,**C**).

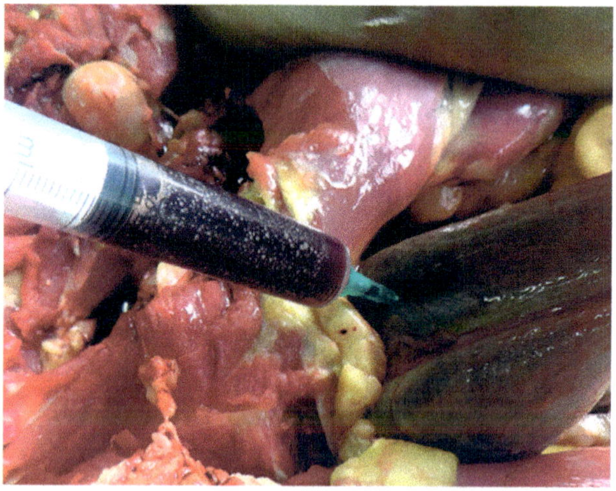

Figure 5. Dark-brown urine—hemoglobinuria, likely resulting from excessive intravascular hemolysis.

3. Discussion

Copper intoxication is a deadly condition in sheep and is reported worldwide [7]. There are two types of copper toxicosis described: acute and chronic [1,7,9,14]. Copper sulfate is typically the form most frequently associated with reported cases of poisoning [9]. In sheep, especially in sensitive breeds like the Texel sheep, the chronic form is more common, and it is usually a consequence of long-term exposure (several weeks) to excessive doses (higher than the dietary requirement) [7]. The incidence of chronic copper poisoning is increasing worldwide as intensive breeding systems are extensively used and sensitive breeds become more popular [7]. The main cause of chronic copper poisoning in sheep is the inability of this species to increase biliary excretion as a response to high dietary intake [14]. High dietary intake will lead to increased absorption, as copper is not absorbed based on the animal's daily requirements like other metals but is dependent and proportionate on the concentration in their diet [15].

In sheep, copper has a complex interrelationship with several elements, especially molybdenum and sulfur [9], as both elements decrease copper absorption and inhibit copper utilization. These minerals interact to form soluble thiomolybdates (mono, di-, tri- and tetra) in the rumen [1]. Thiomolybdates are also absorbed into the bloodstream and act as a copper chelator, keeping copper in a biologically non-functional form. Molybdenum plays a major role, with low dietary levels of the metal being associated with increased absorption of copper. Other metals and minerals that interact with copper absorption are interacting zinc and iron [15]. Interaction with the mentioned minerals, even if the dietary level of copper is within the normal range, may also lead to the development of chronic poisoning [1,7,9].

The elevated copper level in the diet (or molybdenum deficiency) leads to liver accumulation of the metal [1]. The chemical form of copper in the diet is also important as the bioavailability may vary among different products. Newer feed additives, such as copper bilysinate have high bioavailability [7]. In addition to the high bioavailability, the commercial product involved in this episode reduces copper excretion [13].

While the liver accumulates copper, the plasma and serum copper concentrations are maintained within the normal range [1]. When copper levels in the liver reach high, toxic levels, and the storing capacity of lysosomes is overwhelmed, liver necrosis occurs [14]. Copper is then released from the necrotized hepatocytes into the bloodstream, causing oxidation of the erythrocyte membrane resulting in methemoglobinemia and intravascular hemolysis, hemoglobinuria, and elevated serum levels [9,14]. The copper release from the hepatocytes is usually triggered by a stress such as transportation, pregnancy, lactation, extreme weather, and concurrent diseases [1,7,9].

The cause of death in chronic copper poisoning is either acute anemia due to hemolysis or acute renal tubular necrosis induced by the direct toxic effect of the reabsorbed copper by the renal tubular epithelial cells and by the toxic effects of hemoglobin, which is reabsorbed after the hemolytic event [7,9,14].

In the present episode of chronic copper poisoning, the diagnosis was based on the history, clinical data, and post-mortem findings. Feeding a susceptible breed with a high copper diet over several weeks led to copper accumulation in the liver and release of the metal with typical clinical signs and lesions for chronic copper intoxication. Previous studies [16] showed that copper lysine complex administration leads to higher liver concentrations compared to copper sulfate in pigs [16]. In this study, the use of a commercial compound feed containing copper bilysinate, which, according to the manufacturer, reduces copper excretion by more than 43% in the target species (pigs), led to high liver concentrations, hepatocellular necrosis and intravascular hemolysis in the sheep involved in this episode.

Breed is one of the most important factors influencing copper toxicity, and genetic factors could be involved [7]. Merino sheep are less often affected by copper toxicoses than other sheep breeds [17]. North Ronaldsay, reported to be the most susceptible breed, is a primitive sheep adapted to the copper-deprived North Ronaldsay island [7,17]. Finnsheep seems to be the most resistant [18]. A copper content of less than 15 ppm in the diet is considered safe yet it can accumulate in the liver of Texel sheep and Texel x Friesian milk sheep at a relatively high level, which suggests that the sensibility of these breeds to copper is high [18]. Breeds susceptible to copper intoxication, like Texel, along with mature British breeds, like Suffolks, Oxfords, and Shropshires, are often used in the sheep industry, increasing the incidence of the disease [6,7,18].

As the clinical signs in chronic copper poisoning are associated with an acute hemolytic crisis, the differential diagnosis should exclude several infectious diseases such as babesiosis, leptospirosis, theileriosis, anaplasmosis, bacillary hemoglobinuria, Clostridium perfringens type A infection, and toxic plant ingestion such as rape, kale, onion, garlic [7].

Animals suffering from chronic copper poisoning have a poor prognosis and can be difficult to treat when they show severe clinical signs [7] such as in the present episode, where even after the compound feed was removed, another animal died. It is therefore recommended that animals with copper poisoning are treated before the onset of severe clinical signs [9].

Treatment of chronic copper poisoning is realized by administering copper antagonists and chelating agents [7,9,14]. In animals with severe anemia following intravascular hemolysis, a blood transfusion should be considered [7,9]. The most used antagonists are ammonium or sodium molybdate and sodium thiosulphate. These products are administered as a drench, daily, orally, 50–500 mg ammonium or sodium molybdate and 0.3–1 g sodium thiosulphate. The therapy should last 3–4 weeks [7,9,14].

The most effective product in reducing liver copper levels and reducing mortality rates is ammonium tetrathiomolybdate. It is administered intravenously, at 2.7 mg/kg for 3–6 treatments (on alternate days) or subcutaneously at 3.4 mg/kg on three alternate days [7,9,14].

The chelator used in chronic copper poisoning in sheep is D-penicillamine 10 to 15 mg/kg orally twice daily [14]. Sodium calcium edetate and 2,3-dimercapto-1-propanol have also been experimentally used to treat copper poisoning in sheep with variable success [7,14].

Preventing hepatic copper accumulation and reducing the risk of chronic poisoning in susceptible breeds, especially in intensive breeding systems or endemic areas, are mainstays of ovine health management [7]. Prevention is realized with the addition of copper antagonists to the diet. Dietary supplementation with molybdenum (can be increased to 5 ppm in the diet) and sulfur (0.35% of diet) reduces hepatic Cu accumulation in sheep. Zinc supplementation at 100 ppm will also reduce copper absorption [6,14]. A copper/molybdenum ratio of 6:1 to 10:1 in the diet decreases the likelihood of copper accumulation in the liver [14].

Copper bilysinate is a relatively new, recently registered supplement. To the authors' knowledge, this study is the first report of poisoning in sheep involving this new form of copper. In addition to exposure to this highly bioavailable copper source, other ingredients of the commercial compound feed involved in this episode may have contributed to reduced excretion of copper, finally leading to accumulation and chronic copper poisoning. It is essential not to use compound feeds intended for copper-tolerant species like pigs in the diet of sheep.

Author Contributions: Conceptualization, D.P. and A.-L.N.; writing—original draft preparation, D.P.; writing—review and editing, A.-L.N. and A.N.O.; investigation and validation, A.T.; formal analysis and visualization, P.B.; supervision, F.C. All authors have read and agreed to the published version of the manuscript.

Funding: This research received no external funding.

Institutional Review Board Statement: Not applicable.

Informed Consent Statement: Not applicable.

Data Availability Statement: The data generated in this study are presented in this article.

Conflicts of Interest: The authors declare no conflicts of interest.

References

1. Ensley, S. Evaluating Mineral Status in Ruminant Livestock. *Vet. Clin. N. Am. Food Anim.* **2020**, *36*, 525–546. [CrossRef] [PubMed]
2. Andreini, C.; Banci, L.; Bertini, I.; Rosato, A. Occurrence of copper proteins through the three domains of life: A bioinformatic approach. *J. Proteome Res.* **2008**, *7*, 209–216. [CrossRef] [PubMed]
3. Linder, M.C.; Hazegh-Azam, M. Copper biochemistry and molecular biology. *Am. J. Clin. Nutr.* **1996**, *63*, 797S–811S. [CrossRef] [PubMed]
4. Linder, M.C. *Biochemistry of Copper*, 1st ed.; Springer: New York, NY, USA, 1991; Volume 526, pp. 1–13. [CrossRef]
5. Nose, Y.; Wood, L.K.; Kim, B.; Prohaska, J.R.; Fry, R.S.; Spears, J.W.; Thiele, D.J. Ctr1 an apical copper transporter in mammalian intestinal epithelial cells in vitro that is controlled at the level of protein stability. *J. Biol. Chem.* **2010**, *285*, 32385–32392. [CrossRef] [PubMed]
6. Gupta, R.K. A review of copper poisoning in animals: Sheep, goat and cattle. *Int. J. Vet. Sci. Anim. Husb.* **2018**, *3*, 1–4.
7. Borobia, M.; Villanueva-Saz, S.; Ruiz de Arcaute, M.; Fernández, A.; Verde, M.T.; González, J.M.; Navarro, T.; Benito, A.A.; Arnal, J.L.; De las Heras, M.; et al. Copper Poisoning, a Deadly Hazard for Sheep. *Animals* **2022**, *12*, 2388. [CrossRef] [PubMed]
8. Blakley, B.R. Copper Poisoning in Animals. Available online: https://www.merckvetmanual.com/toxicology/copper-poisoning/copper-poisoning-in-animals (accessed on 18 October 2024).
9. Thompson, L.J. Copper. In *Veterinary Toxicology*, 3rd ed.; Gupta, R.C., Ed.; Academic Press (Elsevier): New York, NY, USA, 2018; pp. 425–427. [CrossRef]
10. Bremner, I. Manifestations of copper excess. *Am. J. Clin. Nutr.* **1998**, *67* (Suppl. 5), 1069S–1073S. [CrossRef] [PubMed]
11. Caloni, F.; Berny, P.; Croubels, S.; Sachana, M.; Guitart, R. Epidemiology of Animal Poisonings in Europe. In *Veterinary Toxicology*, 3rd ed.; Gupta, R.C., Ed.; Academic Press (Elsevier): New York, NY, USA, 2018; pp. 45–56. [CrossRef]
12. Commission Implementing Regulation (EU) No 1230/2014 of 17 November 2014 Concerning the Authorisation of Copper Bilysinate as a Feed Additive for All Animal Species. Available online: www.legislation.gov.uk/eur/2014/1230 (accessed on 20 May 2024).
13. NATUPIG Concept: LfL Bavaria Achieves Convincing Test Results in Pig Fattening. Available online: https://www.schaumann.info/en/products-in-field-tests-243/c/natupig-concept-lfl-bavaria-achieves-convincing-test-results-in-pig-fattening-3478 (accessed on 18 September 2024).
14. Moeller, R.B. Copper. In *Clinical Veterinary Toxicology*, 1st ed.; Plumlee, K.H., Ed.; Mosby: St. Louis, MO, USA, 2004; pp. 195–197.
15. Copper Toxicity in Sheep. Ontario Veterinary College. Available online: https://ovc.uoguelph.ca/ruminant_health_management/copper-toxicity-sheep (accessed on 15 April 2024).
16. Apgar, G.A.; Kornegay, E.T.; Lindemann, M.D.; Notter, D.R. Evaluation of copper sulfate and a copper lysine complex as growth promoters for weanling swine. *J. Anim. Sci.* **1995**, *73*, 2640–2646. [CrossRef] [PubMed]
17. Wiener, G.; Wilmut, I.; Field, A.C. Maternal and lamb breed interactions in the concentration of copper in tissues and plasma of sheep. In *Trace Element Metabolism in Man and Animals—3*; Kirchgessner, M., Ed.; Technische Universitat Munchen: Freising-Weihenstephan, Germany, 1978; pp. 469–472.
18. van der Berg, R.; Levels, F.H.; van der Schee, W. Breed differences in sheep with respect to the accumulation of copper in the liver. *Vet. Q.* **1983**, *5*, 171–174. [CrossRef] [PubMed]

MDPI AG
Grosspeteranlage 5
4052 Basel
Switzerland
Tel.: +41 61 683 77 34

Life Editorial Office
E-mail: life@mdpi.com
www.mdpi.com/journal/life

www.ingramcontent.com/pod-product-compliance
Lightning Source LLC
Chambersburg PA
CBHW062034220926
51850CB00010B/2547
* 9 7 8 3 7 2 5 8 5 8 7 3 6 *